Deep Neural Networks-Enabled Intelligent Fault Diagnosis of Mechanical Systems

This book aims to highlight the potential of deep learning (DL)-enabled methods in intelligent fault diagnosis (IFD), along with their benefits and contributions.

The authors first introduce basic applications of DL-enabled IFD, including autoencoders, deep belief networks, and convolutional neural networks. Advanced topics of DL-enabled IFD are also explored, such as data augmentation, multi-sensor fusion, unsupervised deep transfer learning, neural architecture search, self-supervised learning, and reinforcement learning. Aiming to revolutionize the nature of IFD, this book contributes to improved efficiency, safety, and reliability of mechanical systems in various industrial domains.

The book will appeal to academic researchers, practitioners, and students in the fields of intelligent fault diagnosis, prognostics and health management, and deep learning.

Ruqiang Yan is a professor at the School of Mechanical Engineering, Xi'an Jiaotong University. His research interests include data analytics, artificial intelligence, and energy-efficient sensing and sensor networks for the condition monitoring and health diagnosis of large-scale, complex, dynamical systems.

Zhibin Zhao is an assistant professor at the School of Mechanical Engineering, Xi'an Jiaotong University. His research interests include sparse signal processing and machine learning, especially deep learning for machine fault detection, diagnosis, and prognosis.

Deep Neural Networks-Enabled Intelligent Fault Diagnosis of Mechanical Systems

Ruqiang Yan and Zhibin Zhao

CRC Press is an imprint of the
Taylor & Francis Group, an **informa** business

Designed cover image: © AlpakaVideo

This book is published with financial support from the Natural Science Foundation of China under Grant 51835009 and Grant 52105116

First edition published 2024
by CRC Press
2385 NW Executive Center Drive, Suite 320, Boca Raton FL 33431

and by CRC Press
4 Park Square, Milton Park, Abingdon, Oxon, OX14 4RN

CRC Press is an imprint of Taylor & Francis Group, LLC

ISBN: 978-1-032-75237-2 (hbk)
ISBN: 978-1-032-75724-7 (pbk)
ISBN: 978-1-003-47446-3 (ebk)

DOI: 10.1201/9781003474463

Typeset in Minion
by KnowledgeWorks Global Ltd.

Contents

CHAPTER 1

Introduction and Background

INTRODUCTION

In recent years, due to the rapid development of computer technology, modern testing technology, and signal processing technology, equipment fault diagnosis technology has made great progress. With the rapid development of artificial intelligence technology, the application of deep neural network (DNN) in intelligent fault diagnosis (IFD) of mechanical systems has further deepened.

Deep learning (DL) is one of the hottest technologies in the current field of machine learning, and the *MIT Technology Review* ranked DL at the top of the top ten breakthrough technologies of 2013 (Cohen 2015). DL is essentially a DNN with multiple hidden layers, and the main difference between it and the traditional multi-layer perceptron is the difference in the learning algorithm. In 2006, Professor Hinton of the University of Toronto, a leader in the field of machine learning, first proposed the concept of "deep learning" in an article published in *Science* magazine, thus opening the wave of DL research (Hinton and Salakhutdinov 2006). In 2015, a review published in *Nature* stated that DL allows computational models, composed of multiple processing layers, to learn data representations with multiple levels of the abstraction (LeCun, Bengio, and Hinton 2015).

Due to its strong representation learning ability, DL is well-suited for data analysis and classification. Therefore, in the field of intelligent diagnosis, many researchers have applied DL-based techniques, such as multi-layer perceptron (MLP), autoencoder (AE), convolutional neural networks (CNNs), deep belief networks (DBNs), and recurrent neural networks (RNNs), to boost the performance. However, different researchers often recommended to use different inputs (such as time domain input, frequency domain input, time–frequency domain input, etc.) and set different hyper-parameters (such as the learning rate, the batch size, the network architecture, etc.).

Important conferences in the field of neural computing and machine learning, Conference on Neural Information Processing Systems (NIPS) and International Conference on Machine Learning (ICML), have set up DL topics since 2011, and companies such as Google, Microsoft, and Baidu have established special DL R&D departments to conduct industrialized research

DOI: 10.1201/9781003474463-1

on this technology. It can be seen that DL technology has become an emerging research hotspot in academia and industry.

PROGNOSTICS AND HEALTH MANAGEMENT FOR MECHANICAL SYSTEMS

The main focus of this section is to discuss the definition and development of PHM, to present a review that includes developed and applied PHM methods, and to present several unresolved challenges related to PHM.

Introduction of PHM

PHM, including monitoring, diagnosis, prognosis, and health management, occupies an increasingly important position in reducing costly breakdowns and avoiding catastrophic accidents in modern industry. The relationship among these is summarized in Figure 1.1.

Monitoring refers to fault detection, and its purpose is to determine whether the system is in a normal operating state, in which anomaly detection is one of the most important tools to trace the corresponding health state. Diagnosis refers to the identification of the fault type and its corresponding degree. Prognosis makes use of appropriate models to assess the degree of performance degradation and further predicts the remaining useful life (RUL). Health management integrates outputs from monitoring, diagnosis, and prognosis, and makes optimal maintenance and logistic decisions considering economic costs and other available resources.

An effective PHM system is expected to provide early detection and isolation of precursors and/or components or sub-elements of early failures; have the means to monitor and predict the progression of failures; assist in the development or autonomous triggering of maintenance programs and asset management decisions or actions. The condition of detected early failures should be monitored, progressing from a small failure to a larger one, until some maintenance action and/or replacement is required. By using such a system, the health of a machine, component, or system can be known at any point in time and the eventual occurrence of a failure can be predicted and prevented, enabling near-zero

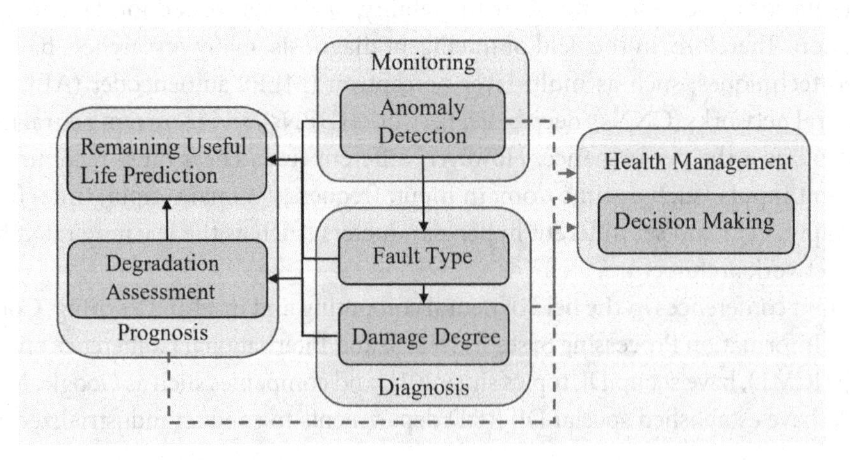

FIGURE 1.1 Relationship between monitoring, diagnosis, prognosis, and health management.

downtime performance. Unnecessary and costly preventive maintenance can be eliminated, maintenance schedules can be optimized, and delivery times for spare parts and resources can be reduced—all of which can lead to significant cost savings. In general, PHM will greatly improve the operational safety, system reliability, and maintainability of equipment, and reduce the cost of equipment throughout its life cycle at the same time.

Development of PHM

The concepts and frameworks of PHM are based on known maintenance methods and diagnostic techniques such as preventive maintenance (PM), reliability-centered maintenance (RCM), and condition-based maintenance (CBM) (Lee, Wu, and Zhao 2014). The future development of PHM will be mutually inspired and facilitated from various fields in addition to engineering.

CBM comprises data acquisition and data processing (condition monitoring) to generate actionable condition information to inform maintenance decisions, thus avoiding unnecessary maintenance tasks. In recent years, more and more research efforts have turned to predictive and health management, i.e., early failure detection, current health assessment, and RUL prediction. However, maintenance strategies should be different in various maintenance scenarios with different system complexity and uncertainty. Figures 1.1 and 1.2 show diagrams of the conversion relationship between CBM, RCM, and PHM, which show the different maintenance strategies with the complexity and uncertainty of the system.

CBM can be applied to systems that (1) can be considered deterministic to some extent, (2) are smooth or stationary, (3) and for which signaling variables can be extracted as indicators of good health, albeit with low dimensionality. If the system is a probabilistic system whose outputs cannot be easily determined using known relational models and are conditionally dependent on the inputs, then the output data will not always be repeatable when observed at different times. As a result, future behavior cannot be accurately predicted based on the domain knowledge and historical observations of the system. For such a system full of uncertainty, RCM is more suitable. RCM focuses on the system's ability to have expected reliability over time and utilizes statistical tools such as failure modes and effects

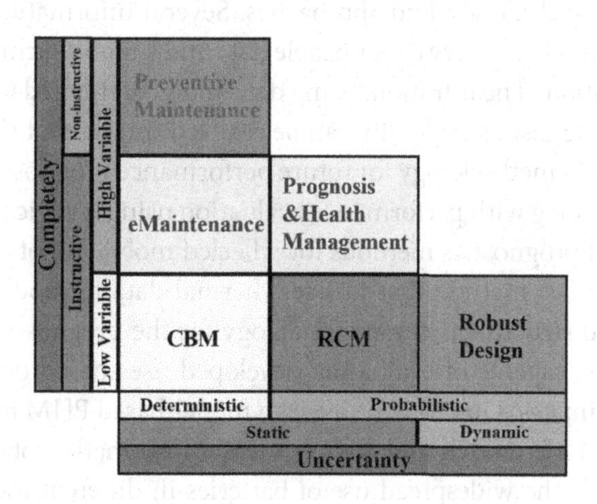

FIGURE 1.2 Maintenance transformation map.

criticality analysis (FMECA) to retrieve information that can help identify failure modes and the probable duration before each mode occurs. Since RCM relies on statistical estimates of total operational life expectancy, it can reduce unplanned or unnecessary maintenance if the system is static and failure modes are well studied. However, RCM is still prone to large deviations in system dynamics and lacks insight into actual system performance. Robust design should be considered if the uncertainty of the system is more complex, such as in a highly dynamic system whose behavior changes over time.

PHM can be viewed as an evolutionary form of CBM. CBM technologies can provide inputs to predictive models in PHM and support timely and accurate decision-making to prevent downtime and maximize profits. Because of its ability to assess health and predict failures and downtime, PHM, when complemented with other technologies, is considered the foundation for advanced fields including self-maintenance, resilient systems, and engineered immune systems. The discipline of PHM needs to be further developed and expanded to help build these fields.

Developed and Applied PHM Methods

Current prognostic methods can be categorized into three groups, namely model-based methods, data-driven methods, and hybrid prognostic methods. A typical model-based approach consists of data generated from simulation models under nominal and degraded conditions. Predictions of the remaining life of the system are generated by simulating multiple modes of operation of the system and blending the predictions of each mode with time-averaged model probability weights. If a reliable or accurate model of the system is not available, data-driven prognostic method is used to determine the remaining lifetime by trending the trajectory of developing failures and predicting the amount of time before a predetermined threshold is reached. Two well-known tracking and prediction tools, the alpha-beta-gamma tracking filter and the Kalman filter, have been applied to gearbox prediction. Both filters have been investigated for their ability to track and smooth the features of gearbox vibration data.

The concept of PHM has been deployed in several applications, leading to the development of various tools, techniques, and approaches. Several information fusion systems for engine PHM that integrate a variety of applicable data and suitable techniques are presented.

By using a combination of health-monitoring data and model-based techniques, a comprehensive component prognostics capability can be realized throughout the component lifecycle. Literature proposes a methodology for future performance prognostics of liftgate systems using ARMA models along with performance evaluation using logistic regression. Literature reviewed model-based prognostics methods for wheeled mobile robots. Literature describes an electronics prognostics method that utilizes thermal data to model stress and damage to electronic parts and structures. In a methodology for the prognostics of electronics, it is demonstrated that it is capable of evaluating developed uses based on phase growth rates and remaining life estimates. Literature proposes a model-based PHM technique for avionics that utilizes embedded life models and environmental information obtained from aircraft-mounted sensors. With the widespread use of batteries in different industries, data-driven battery-health-monitoring methods have emerged. These methods utilize voltage, current, temperature, and other measurements and use Bayesian techniques, such as particle filtering, to estimate battery RUL by combining the collected measurements with system dynamics. A number of research activities have also been reported on the structure of forecasting systems, such as maintenance levels, inventory policies, and supply chain management.

Challenges of PHM

Due to the complexity of the physics involved in PHM in practice, the data available, and the requirements of real-world applications for PHM solutions, a number of challenges remain (Zhao et al. 2021).

Physics of the Problem

The complexity, dynamics, and high nonlinearity of the structures, systems and components (SSCs) degradation process makes it difficult to understand, characterize, and model the physical process of SSCs degradation.

Data Available

The challenges related to the data come from multiple aspects: the many anomalies in the real data collected in the field; the scarcity and incompleteness of the data in relation to the degraded state of the SSCs of interest; the difficulty in managing and processing big data because of the large number of signals collected by the different types of sensors; and the changing operational and environmental conditions that can affect the data used for the training of the PHM models and the calibration of their parameters, as well as the data to which these models are applied.

Requirements of the Solutions

The challenges related to the requirements of the PHM solutions come from the multiple objectives that they must achieve, depending on the applications. The obvious ones are accuracy and precision, based on the vision that defines the performance metrics and the measurements they support: in some cases very high accuracy and precision are needed to be able to take confident decisions, in other cases accuracy and precision don't have to be as high and may compromise other goals.

INTELLIGENT FAULT DIAGNOSIS FOR MECHANICAL SYSTEMS

This section provides background and development of IFD.

Introduction of IFD

IFD is the application of machine learning theory in machinery fault diagnosis (Lei, Yang, and Jiang 2020). It is an effective method to reduce human labor and automatically identify the health status of machines.

Mechanical equipment is subject to performance degradation and even failure during operations. Therefore, in addition to health monitoring of equipment, it is necessary to diagnose mechanical equipment failures. Failure information of equipment is often implicit in monitoring data such as vibration, acoustic emission, and temperature.

Effectively capturing the fault information conveyed by the monitoring data changes, you can accurately identify the location, type, and even degree of failure of mechanical equipment. Long-term experience enables engineers to capture the health status information in the monitoring data changes. Therefore, by utilizing the empirical knowledge, it is possible to make a judgment on the health status of mechanical equipment. However, with the rapid improvement of the collection, storage, and transmission capabilities of the sensing system, the number, size, and type of monitoring signals of machinery and equipment continue to increase, providing a huge amount of data for machinery and equipment fault diagnosis, and it is difficult to satisfy the demand for fault diagnosis driven by big data only by the accumulation of human experience.

With the rapid development of artificial intelligence, machine learning technology tries to give computers the ability to learn, so that they can analyze the data, generalize the laws, summarize the experience, and ultimately replace the human learning or their own experience accumulation process, and liberate human beings from the complicated sea of data.

Therefore, IFD can be understood as the science of using big data to identify the health status of mechanical equipment, that is, based on the monitoring data obtained by the sensing system and through the accumulated experience and knowledge of machine learning, the intelligent recognition of the health status of the equipment ensures the reliability of mechanical equipment operation.

Machine Learning-Enabled IFD

Most of the traditional fault diagnosis processes are performed by manually checking the health of the machine, which increases the labor intensity and reduces the accuracy of the diagnosis. Advanced signal processing methods can help determine the types of faults or where in the machine the faults are occurring. However, these methods rely heavily on the expertise that maintenance personnel mostly lack in engineering scenarios. Moreover, the diagnosis results of signal processing methods are too specialized to be understood by machine users. As a result, modern industrial applications prefer fault diagnosis methods that can automatically recognize the health of a machine. With the help of machine learning theory, IFD is expected to fulfill the above purpose. In past IFD research, some traditional machine learning theories, such as artificial neural networks and support vector machines, were applied to machinery fault diagnosis. The diagnosis procedure consists of three steps: data collection, artificial feature extraction, and health state identification, each of which is described in detail in the following subsections.

Data Collection

In the data collection step, sensors are mounted on the machine to continuously collect data. Different sensors such as vibration sensors, acoustic emission sensors, temperature sensors, and current transformers are usually used. Among them, vibration data are widely used for troubleshooting bearings and gearboxes. Acoustic emission data has the potential to detect early failures and deformations in bearings and gears, especially in low-speed operation and low-frequency noise environments. Instantaneous RPM data is commonly used in engine troubleshooting and is highly resistant to interference. Current data plays an important role in motor troubleshooting. This data can be easily collected using only current transformers and is not included in the operation of the machine. In addition, researchers have found that data from multiple sources of sensors have complementary information that can be fused together to achieve higher diagnosis accuracy than data from just a single sensor.

Artificial Feature Extraction

Artificial feature extraction is a two-step process. First, some commonly used features such as time domain features, frequency domain features, and time–frequency domain features are extracted from the collected data. These features contain the operating condition information that reflects the operating condition of the machine. Second, feature selection methods such as filters, wrappers and embedding methods are used to select sensitive features of machine health status from the extracted features. This facilitates the elimination of redundant information and further improves the diagnosis results.

Health State Recognition

Health state recognition uses a machine learning-based diagnosis model to establish the relationship between the selected features and the health state of the machine. To achieve this, the diagnosis model is first trained with labeled samples. The model is then able to recognize the health state of the machine when the input samples are unlabeled.

Expert System-Based Approaches The expert system-based diagnosis models represent the diagnosis knowledge from experts in the form of the inference algorithm to automatically recognize the health states of machines. However, the performance of diagnostic models relies heavily on expert knowledge that is difficult to access and express. Incorrect and incomplete knowledge may reduce the accuracy of diagnosis. In addition, the expert system lacks self-learning capabilities, and the diagnostic knowledge base is difficult to expand and correct.

ANN-Based Approaches Artificial neural networks (ANNs) imitate the activities of human brains for information processing, which is an effective way to establish the diagnosis models. This section reviews the applications of ANNs in fault diagnosis of machines. ANN-based diagnosis models have strong self-learning capabilities and can automatically learn diagnostic knowledge from input data to minimize empirical risks. In addition, they can easily recognize multiple states of a machine. However, there are two drawbacks. First, the complexity of the diagnostic model greatly increases with the increase of monitoring data inputs. The increase in model parameters reduces the training efficiency, which in turn leads to overfitting and reduces the diagnostic accuracy of the diagnostic model. Second, due to the lack of rigorous theoretical support, ANN-based diagnosis models are black-box. Therefore, their interpretability is low.

SVM-Based Approaches Support vector machine (SVM) is a supervised learning method which is widely concerned in classification tasks. We briefly review the theory of SVM and summarize its applications in machine fault diagnosis in this subsection. Different from the

ANN, SVM-based diagnostic models are trained by minimizing the structural risk, which contributes to improving the interpretability of the model due to its theoretical rigor. The optimization objective solution of SVM is convex quadratic optimization which enables the diagnostic model to easily obtain the global optimal solution, which in turn leads to a high diagnostic accuracy. The diagnostic model based on SVM has three drawbacks to consider. First, the diagnostic models can effectively handle a small amount of monitoring data. However, they are difficult to fit to massive amounts of data, which may lead to a computational curse. Second, the performance of SVM-based diagnostic models is very sensitive to kernel parameters. Inappropriate kernel parameters do not even yield reliable diagnostic results. For multi-class classification tasks in IFD, it is often necessary to use complex architectures such as OAA and OAO, etc. to integrate the results of multiple SVM-based models.

Deep Learning-Enabled IFD

With the rapid growth of Internet technology and the Internet of Things (IoT), the amount of data being collected is greater than ever before. The growing amount of data provides more information for machine troubleshooting, which in turn is more likely to provide accurate diagnosis results. Unfortunately, previous fault diagnosis based on traditional machine learning theories is not suitable for such big data scenarios. Therefore, it is necessary to develop some advanced IFD methods. DL originates from the study of neural networks, which uses a deep hierarchy to automatically represent abstract features and further establishes the relationship between the learned features and the target output directly. The diagnosis process based on DL mainly consists of two steps: big data collection and DL-based diagnosis.

Big Data Collection

Big data has become a popular term in modern industrial and other application scenarios. Typically, big data consists of four characteristics, namely volume, velocity, diversity, and accuracy. In contrast, big data for monitoring and controlling machines possesses these characteristics and further extends to very specialized areas. The characteristics are summarized as follows.

Large Amount of Data The amount of data collected will continue to grow over the long-term operation of the machine, especially for large units such as wind turbines in wind farms.

Low Value Density There is incomplete health information in the collected big data. In addition, a certain percentage of poor-quality data is mixed in the massive data.

Multi-Source Heterogeneous Data Structure Multi-source data are collected by different types of sensors. In addition, the data is heterogeneous due to different storage structures.

Monitoring Data Streams High-speed transmission channels enable immediate collection of data from machines.

These characteristics are due to the following conditions. First, in modern industry, most production activities are performed by a group of machines. As a result, troubleshooting tends to be concentrated on the unit and there is a tendency to increase the amount of data collected. Second, although monitoring systems can capture a large amount of data, only a few of them are valuable. In addition, there are many monitoring points on the machine in order to collect enough information about the health of the machine. Finally, the boom in sensor technology and data transmission, especially with the advent of the IoT and high-speed Internet, has led to the collection of large amounts of data containing real-time information.

Deep Learning-Based Diagnosis

DL-based diagnosis models automatically learn features from the input monitoring data and simultaneously identify the health status of the machine based on the learned features. They mainly consist of a feature extraction layer and a classification layer. The model first learns abstract features layer by layer using hierarchical networks such as stacked AE, DBN, CNN, and ResNet. In addition, the output layer is placed after the last extraction layer for health state recognition, usually using a neural network-based classifier because of its high multi-class classification capability. During the training process, the training parameters of the diagnosis model are updated using the BP algorithm to minimize the error between the actual output and the target.

Stacked AE-Based Approaches Stacked AE-based diagnosis models automatically present health information from the input monitoring data. This eliminates the need to rely on much expert knowledge in feature extraction. As an unsupervised learning method, stacked AE cannot be directly used to recognize the health state of a machine. Therefore, a classification layer is usually added to the top layer of the model architecture to recognize the health state of the machine. Therefore, a classification layer is usually added at the top of the model architecture, and the constructed diagnosis model needs to be trained with a sufficiently large number of labeled samples.

DBN-Based Approaches Unlike stacked AE, DBN-based diagnosis models can automatically learn features from input data by pre-training a set of stacked RBMs. This solves the problem of gradient vanishing when using BP algorithms to fine-tune deep networks. To identify the health of the machine, DBN maps the learned features into the label space by adding a classification layer. It is necessary to train the constructed diagnosis model with enough labeled data to obtain convincing diagnosis results.

CNN-Based Approaches Compared with stacked AE and DBN, CNN-based diagnosis models are able to learn features directly from raw monitoring data without preprocessing such as frequency domain transformations, because CNNs are able to capture the shift variant characteristics of the input data. In addition, the number of training parameters in the diagnosis model is reduced through weight assignment, which speeds up convergence and suppresses overfitting. Similar to other DL-based models, the diagnosis performance of CNN-based diagnosis models is limited by the training with sufficient labeled samples.

ResNet-Based Approaches ResNet is developed on the basis of CNN architecture to obtain higher generalization performance. Therefore, the diagnosis model based on ResNet inherits the advantages of the CNN-based diagnosis model and has the potential to obtain higher diagnosis accuracy, especially for the diagnosis of complex working conditions such as variable speed and variable load.

Transfer Learning-Enabled IFD

The success of IFD relies heavily on sufficient labeled data to train machine learning-based diagnosis models. However, recollecting enough data and further labeling them requires significant cost, which is unrealistic for machines in engineering scenarios. Diagnosis knowledge can be reused across multiple related machines, thus solving this problem. For example, diagnosis knowledge from a bearing used in a laboratory can help identify the health of the bearing in an engineering scenario. In this case, it is possible to simulate different failures and collect sufficient marker data from the lab-used bearing. If the diagnosis knowledge can be reused, the diagnosis models trained with them can also be used for diagnosing bearing faults in engineering applications. Migration learning is able to achieve the above purpose, where knowledge from one or more diagnosis tasks can be reused in other related but different diagnosis tasks. With the help of transfer learning theory, there is a need to collect sufficient labeled data, which releases the general assumptions of machine learning-based diagnosis model training. Therefore, IFD is expected to expand from academic research to engineering scenarios.

Transfer Problems in IFD

In transfer learning for IFD, diagnosis knowledge is expected to be reused from one or more diagnosis tasks (source domains) to other related but different diagnosis tasks (target domains), i.e., the concepts of domain and task.

Transfer Scenarios in IFD

Transfer scenarios in IFD can be categorized into two types, i.e., transfers in the same machine (TIM) and transfers across different machines (TDM). Both types of data obey a common assumption that the source domain data is labeled while there is little or no labeled data in the target domain. Migration learning is expected to construct diagnosis models for migration scenarios in IFD.

Categories of Transfer Learning-Based Approaches in IFD

Several researchers have conducted exploratory studies on IFD using transfer learning theory. We further categorize them into four groups, namely, feature-based methods, GAN-based methods, instance-based methods, and parameter-based methods. Among them, feature-based approaches account for the largest proportion of studies. In addition, most of the studies focus on TIM scenarios and only four publications address TDM scenarios.

BASIC CONCEPT OF DEEP NEURAL NETWORKS

Overview

A biological neural network consists of a cluster of neurons that are chemically interconnected or functionally related. A single neuron has the potential to establish connections with numerous other neurons, contributing to a vast network comprising a substantial number of neurons and connections. These connections, referred to as synapses, primarily materialize between axons and dendrites, although other connections such as dendrodendritic synapses are also feasible. In addition to electrical signaling, diverse modes of signaling result from the diffusion of neurotransmitters.

Artificial intelligence, cognitive modeling, and neural networks are information processing paradigms that draw inspiration from the way biological neural systems handle data. Both artificial intelligence and cognitive modeling try to simulate certain characteristics of biological neural networks. Within the realm of artificial intelligence, ANNs have been effectively employed in applications such as speech recognition, image analysis, and adaptive control (Dong, Wang, and Abbas 2021; Minaee et al. 2022). They are utilized to craft software agents, often seen in computer and video games, as well as in the development of autonomous robots. The theory of neural networks has played a crucial role in enhancing our understanding of the functioning of brain neurons, while also forming the foundation for endeavors aimed at generating artificial intelligence (Good fellow, Bengio, and Courville 2016).

Neuron

Firstly, we introduce the structure of neurons in biology. A neuron typically possesses multiple dendrites, primarily for receiving incoming information, while there is only one axon. The axon terminates in numerous axon terminals which can transmit information to multiple other neurons. These axon terminals connect with dendrites of other neurons, thereby transmitting signals. The location of this connection is biologically referred to as a "synapse".

The shape of neurons in the human brain can be illustrated simply as shown in Figure 1.3.

The abstract neuron model is based on the structure of biological neurons. The neuron is a model that includes inputs, outputs, and computational functions. Input can be likened

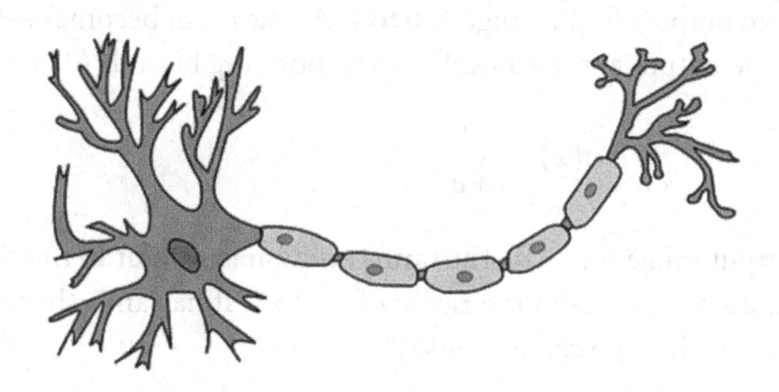

FIGURE 1.3 Diagram of neuron in biology.

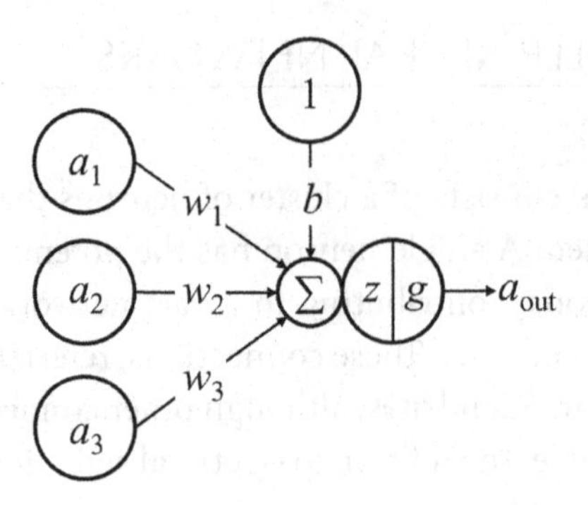

FIGURE 1.4 Diagram of neuron in neural networks.

to the dendrites of a neuron, while output can be compared to the neuron's axon, and the computational function can be likened to the cell nucleus. The neuron model typically includes a bias, which is an additional input to the neuron.

Figure 1.4 represents a typical neuron model that consists of three inputs, one bias, one output, and two computational functions. The arrows between components are referred to as connections, and each connection has an associated weight. The bias always remains as 1 and has its connection weight. This ensures that even if all the other inputs are 0, the neuron will still activate. The training algorithm of a neural network is designed to adjust the values of the weights to their optimum, in order to achieve the best predictive performance for the network.

Assuming a_1, a_2, and a_3 are inputs, w_1, w_2, w_3 and b represent weights, z is the linear combination of inputs and weights, and a_{out} is the result of z after being activated by the function $g(\cdot)$. It can be expressed in formula: $z = a_1w_1 + a_2w_2 + a_3w_3 + b,\ \hat{y} = a_{out} = sigmoid(z) = \frac{1}{1+e^{-z}}$. The function $g(\cdot)$ here is known as an activation function, which is a nonlinear function. Its purpose is to introduce nonlinearity into the output of the neuron. Since real-world data is inherently nonlinear, we want neurons to be able to learn these nonlinear representations. Here are some commonly used activation functions:

- Sigmoid: The output range is [0,1]. The sigmoid function has an S-shaped curve and maps inputs to outputs in the range of 0 to 1. As the input becomes extremely positive or negative, the output approaches the corresponding bounds (0 or 1).

$$\sigma(x) = \frac{1}{1+e^{-x}}$$

- tanh: The output range is [−1,1]. The tanh function is similar to the sigmoid function but maps inputs to outputs in the range of −1 to 1. It has an S-shaped curve like the sigmoid function, but the center is at 0.

$$\tanh(x) = \frac{e^{x} - e^{-x}}{e^{x} + e^{-x}} = 2\sigma(2x) - 1$$

- ReLU (rectified linear unit): The ReLU function is a piecewise linear function that returns the input value for positive inputs and 0 for negative inputs (Glorot, Bordes, and Bengio 2011). It is computationally efficient and has been widely used in DL due to its ability to introduce nonlinearity.

$$f(x)=\max(0,x)$$

The graphs of the aforementioned activation functions are given in Figure 1.5.

Structure of Deep Neural Networks

The simplest neural network is perceptron (Rosenblatt 1958). As shown in Figure 1.4, it consists of only one neuron. Several inputs are combined through linear weighted summation and a nonlinear activation function to produce an output. This model is only suitable for binary classification and cannot learn complex nonlinear models.

Deep Neural Networks (DNNs) are an extension of the perceptron. They incorporate two key enhancements. First, the addition of hidden layers, which can be stacked multiple times, enhancing the model's expressive capabilities. Second, the output layer's neurons can be more than one, allowing for multiple outputs. This flexibility allows the model to be applied effectively to classification, regression, dimensionality reduction, clustering, and other machine learning domains. An example of a DNN is depicted in Figure 1.6. DNNs with linear connections are sometimes referred to as multi-layer perceptron (MLP).

When categorized by the positions of different layers, the layers of a DNN can be divided into three types: input layer, hidden layers, and output layer, as shown in Figure 1.6. Generally, the first layer serves as the input layer, the last layer as the output layer, and the ones in between are the hidden layers. The input layer requires input signals and passes them to the next layer. It performs no operations on the input signals and has no associated weights or biases. In Figure 1.6, there are four input signals. The hidden layers consist of neurons that apply different transformations to the input data. A hidden layer is a collection of vertically arranged neurons. In Figure 1.6, there are five hidden layers. The first hidden layer has four neurons, the second layer has five neurons, the third layer has six neurons, the fourth layer has four neurons, and the fifth layer has three neurons. Each neuron in the hidden layer is connected to every neuron in the next layer, resulting in a fully connected hidden layer. The

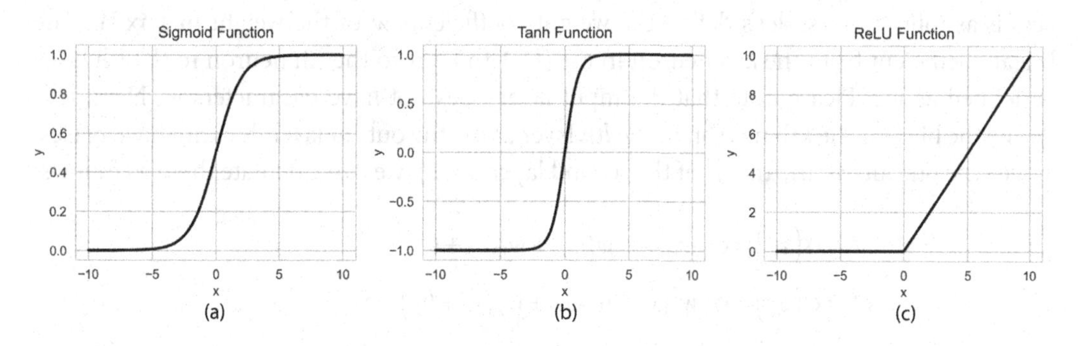

FIGURE 1.5 Activation functions: (a) sigmoid, (b) tanh, and (c) ReLU.

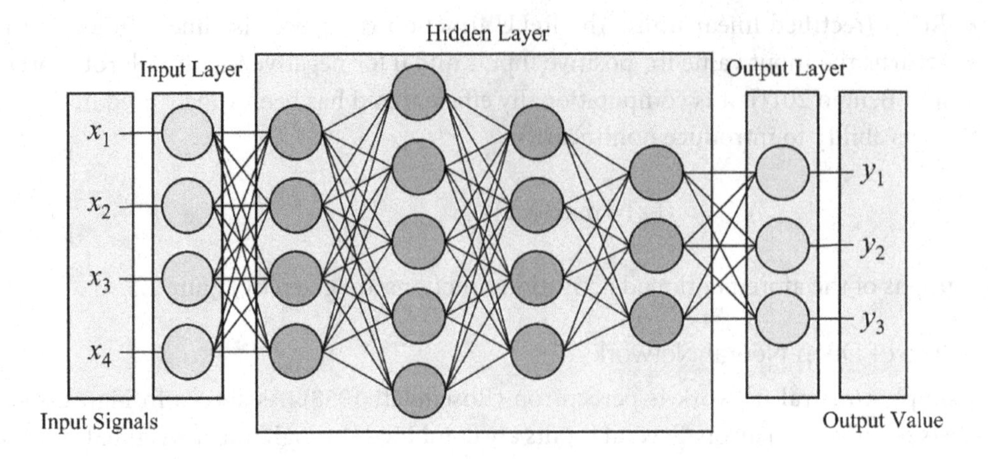

FIGURE 1.6 The structure of a deep neural network.

output layer receives inputs from the last hidden layer. Through this layer, we can determine the expected values and their expected ranges. In Figure 1.6, the output layer has three neurons producing three outputs. Although DNNs might seem complex, they can still be broken down into small local models that resemble perceptron, consisting of a linear relationship $z = \sum w_i x_i + b$ added to an activation function $\sigma(z)$.

Forward Propagation Algorithm

Assuming the chosen activation function is $\sigma(z)$ and the output values of the hidden layers and output layer are represented as a, then for the three-layer DNN shown in Figure 1.5, we can utilize the outputs from the previous layer to compute the outputs of the next layer using a similar approach as the perceptron. This is known as the DNN forward propagation algorithm.

The DNN forward propagation algorithm involves performing a sequence of linear and activation operations using weight matrices W and biases b along with an input value vector x. Starting from the input layer, and following a similar approach as perceptron, this algorithm employs the outputs from the previous layer to calculate the outputs of the next layer. This process continues layer by layer until reaching the output layer, resulting in an output value.

For the three-layer DNN shown in Figure 1.7, the forward propagation calculation process is as follows. First, let's define the weight coefficients w of the weight matrix W. The linear coefficient from the kth neuron in the $(l-1)$th layer to the jth neuron in the lth layer is defined as w_{jk}^l. Please note that the input layer does not have parameters w. Next, let's define the bias for the kth neuron in the lth layer as b_k^l. The output layer does not have biases.

For the outputs a_1^2, a_2^2, and a_3^2 of the second layer $(l=2)$, we can calculate them as follows:

$$a_1^2 = \sigma\left(z_1^2\right) = \sigma\left(w_{11}^2 x_1 + w_{12}^2 x_2 + w_{13}^2 x_3 + b_1^2\right)$$

$$a_2^2 = \sigma\left(z_2^2\right) = \sigma\left(w_{21}^2 x_1 + w_{22}^2 x_2 + w_{23}^2 x_3 + b_2^2\right)$$

$$a_3^2 = \sigma\left(z_3^2\right) = \sigma\left(w_{31}^2 x_1 + w_{32}^2 x_2 + w_{33}^2 x_3 + b_3^2\right)$$

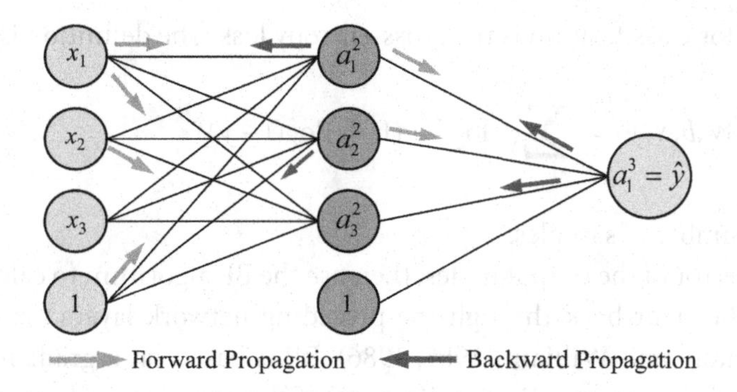

FIGURE 1.7 Forward and backward propagation algorithm of a three-layer deep neural network.

For the output a_1^3 of the third layer ($l=3$), the calculation process is as follows:

$$a_1^3 = \sigma\left(z_1^3\right) = \sigma\left(w_{11}^3 x_1 + w_{12}^3 x_2 + w_{13}^3 x_3 + b_1^3\right)$$

Generalizing the example above, assuming there are m neurons in the $(l-1)$th layer, the calculation for the output a_j^l of the jth neuron in the lth layer can be expressed as follows:

$$a_j^l = \sigma\left(z_j^l\right) = \sigma\left(\sum_{k=1}^{m} w_{jk}^l a_k^{l-1} + b_j^l\right)$$

If $l=2$, $a_k^l = x_k$ in the input layer. In addition, if we use the matrix notation, it becomes more concise. Assuming there are m neurons in the $(l-1)$th layer and n neurons in the lth layer, the linear coefficients w of the lth layer form an $n \times m$ matrix W^l. The biases b of the lth layer form a $n \times 1$ vector b^l. The outputs of the $(l-1)$th layer form a $m \times 1$ vector a^{l-1}. The linear outputs z of the lth layer before activation function form a $n \times 1$ vector z^l. The outputs of the lth layer form a $n \times 1$ vector a^l. Represented using matrix notation, the output of the lth layer is:

$$a^l = \sigma\left(z^l\right) = \sigma\left(Wa^{l-1} + b^l\right)$$

The value of l ranges from 2 to L. The final result of the output layer is a^L.

Backward Propagation Algorithm

Before performing the DNN backpropagation (BP) algorithm, it's necessary to choose a loss function to measure the loss between the calculated output for training samples and the actual output of the training samples. The common method to measure loss is the mean squared error (MSE) which is often used for regression. For each sample, we aim to minimize the following expression:

$$J(W,b,x,y) = \frac{1}{2}\left\|\hat{y} - y\right\|_2^2$$

where $\hat{y}$ is the output at the last layer and has the same shape as y.

The loss used for classification is the cross-entropy loss. The definition is as follows:

$$J(W,b,x,y) = -\sum_{i=1}^{m} (y \log \hat{y} + (1-y)\log(1-\hat{y}))$$

where m is the number of samples.

Compute the error of the output nodes, then use the BP algorithm to calculate gradients and propagate the error back through the preceding network layers for weight updates (Rumelhart, Hinton, and Williams 1985, 1986). BP is based on a gradient descent algorithm, adjusting the parameters in the direction of the negative gradient of the objective, aiming to minimize the error values of the output layer. For the MSE $J(W,b,x,y)$ and given the learning rate η, the update process is expressed as:

$$w_{jk}^{l} \leftarrow w_{jk}^{l} - \eta \frac{\partial J(W,b,x,y)}{\partial w_{jk}^{l}}$$

$$b_{j}^{l} \leftarrow b_{j}^{l} - \eta \frac{\partial J(W,b,x,y)}{\partial b_{j}^{l}}$$

BP uses the chain rule. In the chain rule, we first calculate the derivative of the loss with respect to the corresponding weights in the last layer. Then, we use these gradient values to calculate the gradients in the second-to-last layer. This process is repeated until we obtain gradients for each weight in the neural network.

Forward propagation and BP are repeated continuously, and the neural network weights are iteratively updated until the network converges to the minimum loss.

Regularization

First, let's introduce the concepts of underfitting and overfitting in neural networks. Underfitting occurs when a machine learning model is too simple to capture the underlying patterns in the data. It fails to learn the relationships between input features and the target output, resulting in poor performance on both the training data and unseen data. In other words, the model's capacity is insufficient to represent the complexity of the data. Overfitting happens when a model becomes too complex and starts to memorize the training data instead of learning generalizable patterns. As a result, the model performs very well on the training data but fails to generalize to new, unseen data. Essentially, the model fits the noise in the training data, leading to poor performance in real-world scenarios.

Regularization is a set of techniques often used to prevent or mitigate overfitting. These techniques add additional constraints or penalties to the model's learning process to discourage it from fitting the noise in the training data. The goal is to encourage the model to learn important patterns while avoiding excessive complexity. There are several popular regularization methods:

1. L1 and L2 regularization (Zou and Hastie 2005): These are methods that add a penalty term to the model's loss function based on the magnitude of the model's parameters. L1 regularization adds the absolute values of the parameters, encouraging some of

them to become exactly zero. L2 regularization adds the squared values of the parameters, which discourages large parameter values. A schematic diagram of the L1 regularization and L2 regularization are displayed in Figure 1.8.

2. Dropout: This is a technique commonly used in neural networks (Srivastava et al. 2014). During training, randomly selected neurons are dropped out (i.e., their outputs are set to zero) with a certain probability. This prevents specific neurons from relying too heavily on each other and promotes the generalization of the network.
3. Early stopping: With early stopping, the training process is halted when the model's performance on the validation set starts to degrade. This prevents the model from continuing to train and overfitting the data.

These regularization techniques play a crucial role in achieving a balance between model complexity and generalization ability, ultimately improving a model's performance on unseen data and addressing the issues of underfitting and overfitting.

DL-Enabled IFD

DL, as a booming data mining technique, has swept many fields including computer vision (CV) (Krizhevsky, Sutskever, and Hinton 2012; Farabet et al. 2012), natural language processing (NLP) (Hirschberg and Manning 2015; Sun, Luo, and Chen 2017; Young et al. 2018), and other fields (Feng et al. 2019; D. Li, Liu, and Huang 2020). In 2006, the concept of DL was first introduced via proposing DBNs (Hinton and Salakhutdinov 2006). In 2013, *MIT Technology Review* ranked the DL technology as the top ten breakthrough technologies (Cohen 2015). In 2015, a review (LeCun, Bengio, and Hinton 2015) published in *Nature* stated that DL allows computational models composed of multiple processing layers to learn data representations with multiple levels of abstraction. Due to its strong representation learning ability, DL is well-suited to data analysis and classification. Therefore, in the

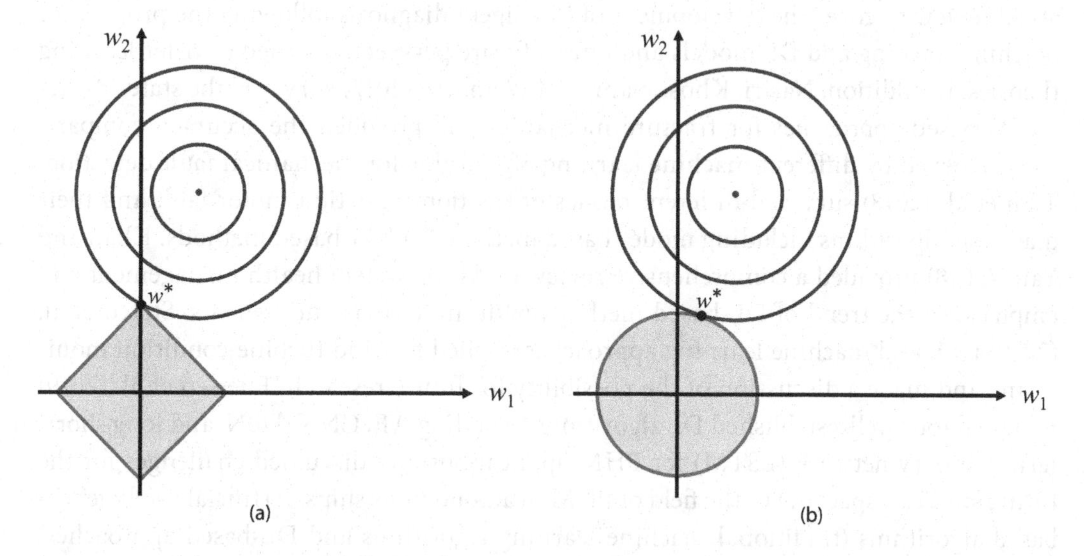

FIGURE 1.8 Diagram of regularization: (a) L1 regularization and (b) L2 regularization.

field of intelligent diagnosis, many researchers have applied DL-based techniques, such as MLP, AE, CNN, DBN, and RNN to boost the performance.

Many review papers on this topic have been published. Therefore, we summarize the main contents of different review papers, allowing readers who just enter this field to find suitable reviews quickly.

In bearing fault diagnosis, Li et al. (2018) provided a systematic review of fuzzy formalisms including combination with other machine learning algorithms. Hoang and Kang (2019) provided a comprehensive review of three popular DL algorithms (AE, DBN, and CNN) for bearing fault diagnosis. Zhang et al. (2020) systematically reviewed the machine learning and DL-based algorithms for bearing fault diagnosis and also provided a comparison of the classification accuracy of CWRU with different DL-based methods. Hamadache et al. (2019) reviewed different fault modes of rolling element bearings and described various health indexes for PHM. Meanwhile, it also provided a survey of artificial intelligence methods for PHM including shallow learning and DL.

In rotating machinery intelligent diagnosis, Ali et al. (2016) provided a review of artificial intelligence-based methods using acoustic emission data for rotating machinery condition monitoring. Liu et al. (2018) reviewed AI-based approaches including k-nearest neighbors (KNN), SVM, ANNs, Naive Bayes, and DL for fault diagnosis of rotating machinery. Wei et al. (2019) summarized early fault diagnosis of gears, bearings, and rotors through signal processing methods (adaptive decomposition methods, wavelet transform, and sparse decomposition) and AI-based methods (KNN, neural network, and SVM). In machinery condition monitoring, Zhao et al. (2016) and Duan et al. (2018) reviewed diagnosis and prognosis of mechanical equipment based on DL algorithms such as DBN and CNN. Zhang et al. (2017) reviewed computational intelligent approaches including ANN, evolutionary algorithms, fuzzy logic, and SVM for machinery fault diagnosis. Zhao et al. (2019) reviewed data-driven machine health monitoring through DL methods (AE, DBN, CNN, and RNN) and provided the data and codes (in Keras) about an experimental study. Lei et al. (2020) presented a systematical review to cover the development of intelligent diagnosis following the progress of machine learning and DL models and offer a future perspective called transfer learning theories. In addition, Nasiri, Khosravani, and Weinberg (2017) surveyed the state-of-the-art AI-based approaches for fracture mechanics and provided the accuracy comparisons achieved by different machine learning algorithms for mechanical fault detection. Tian et al. (2018) surveyed different modes of traction induction motor fault and their diagnosis algorithms including model-based methods and AI-based methods. Khan and Yairi (2018) provided a comprehensive review of AI for system health management and emphasized the trend of DL-based methods with limitations and benefits. Stetco et al. (2019) reviewed machine learning approaches applied to wind turbine condition monitoring and made a discussion of the possibility for future research. Ellefsen et al. (2019) reviewed four well-established DL algorithms including AE, CNN, DBN, and long short-term memory network (LSTM) for PHM applications and discussed challenges for the future studies, especially in the field of PHM in autonomous ships. Artificial intelligence-based algorithms (traditional machine learning algorithms and DL-based approaches) and applications (smart sensors, intelligent manufacturing, PHM, and cyber-physical systems) were reviewed by Ademujimi, Brundage, and Prabhu (2017), Chang, Lee, and Liu (2018), Wang et al. (2018), and Sharp, Ak, and Hedberg Jr (2018) for smart manufacturing and manufacturing diagnosis.

In addition to different designs for the architecture of neural networks such as MLP, DBN, AE, CNN, RNN, and so on, there are some other DL techniques for IFD.

Generative adversarial networks are effective for data augmentation, especially when the quantity of data collected from machines is limited. DL approaches are built upon the foundation of high-quality big data. In cases where training data is insufficient, the network is prone to overfitting, making it challenging to generalize across different scenarios.

Multi-sensor fusion involves the study of integrating measurements from multiple sensors within neural networks. The main idea is to extract the essential features that represent the faults by combining information from various sensors. Fusion can be achieved in three levels: data-level, feature-level, and decision-level. Neural network model with the fusion of multiple sensors exhibits improved accuracy, stability, and the ability to overcome the overfitting problem.

Unsupervised deep transfer learning is a well-known tool for addressing the challenge of having limited labeled data or no labeled data. In the field of IFD, the problem of few labeled fault data in industry and various working conditions is disadvantageous for training a neural network model. Deep transfer learning proves valuable in acquiring transferable features via network-based, instanced-based, mapping-based, and adversarial-based approaches.

The data in different domains of the key components in devices are inconsistent, and can also be easily shifted in changing environment. For this problem, it is natural to design a matching model for data in each domain. However, the architecture parameters of network are complex, discrete, and disordered hyper-parameters, and require to be adjusted from multiple dimensions. It is not feasible to design matching network architecture manually for data in each domain. Neural architecture search (NAS) helps to solve the problem. NAS helps to find the best network architecture for data automatically.

Taking practical scenarios into account, the amount of labeled data is insufficient because of the difficulty of data annotation, which would amplify the potential for overfitting. However, there is a substantial volume of unlabeled data which isn't fully explored. Self-supervised learning integrates the information from the massive unlabeled data and enriches the capacity of learnable data.

Different from supervised learning where the learning ways are static, deep reinforcement learning follows the human cognitive mechanism that knowledge is learned little by little through interaction with the environment. The agent can autonomously learn the behavior policy by interacting with the environment to maximize the cumulative future reward and improve the reliability and generalization ability.

There are more DL methods for IFD, which will not be listed here. Subsequent sections will provide a detailed presentation of the specific applications of those methods in the context of IFD.

The rest of this book has been divided into two parts in total, with Chapters 2 through 4 being the first part, which focuses on basic applications of DL-enabled intelligent fault diagnosis, and Chapters 5 through 10 being the second part, which focuses on advanced topics of DL-enabled intelligent fault diagnosis. In Chapter 2, we focus on AEs for IFD. DBNs for IFD are discussed in Chapter 3. Chapter 4 is about CNNs for IFD. Chapter 5 reviews the data augmentation for IFD. In Chapter 6, we discuss the multi-sensor fusion for IFD. Chapter 7 argues the unsupervised deep transfer learning for IFD. Chapter 8 introduces neural architecture search for IFD. Chapter 9 includes self-supervised learning (SSF) for IFD. Chapter 10 covers reinforcement learning for IFD.

CONCLUSION

To conclude, this chapter lays the groundwork for the subsequent exploration of DL-enabled IFD of mechanical systems. First, it established the significance of PHM and IFD for mechanical systems. Then, it introduced the basic concepts of DNN. A brief review of the techniques employed for DL-enabled IFD was provided. And then it presented a road-map for the subsequent chapters. As we delve deeper into autoencoders, DBNs, CNNs, and advanced topics such as data augmentation, multi-sensor fusion, unsupervised deep transfer learning, neural architecture search, self-supervised learning, and reinforcement learning, we aim to unveil the potential of these methodologies in IFD, bringing to light their strengths and contributions. We strive to revolutionize the essence of IFD, thereby contributing to enhanced efficiency, safety, and reliability of mechanical systems across diverse industrial domains.

REFERENCES

Ademujimi, Toyosi Toriola, Michael P. Brundage, and Vittaldas V. Prabhu. 2017. "A Review of Current Machine Learning Techniques Used in Manufacturing Diagnosis." Advances in Production Management Systems. The Path to Intelligent, Collaborative and Sustainable Manufacturing: IFIP WG 5.7 International Conference, APMS 2017, Hamburg, Germany, September 3–7, 2017, Proceedings, Part I.

Ali, Yasir Hassan, Salah M. Ali, R. A. Rahman, and Raja Ishak Raja Hamzah. 2016. "Acoustic Emission and Artificial Intelligent Methods in Condition Monitoring of Rotating Machine—A Review." National Conference for Postgraduate Research.

Chang, Chih-Wen, Hau-Wei Lee, and Chein-Hung Liu. 2018. "A Review of Artificial Intelligence Algorithms Used for Smart Machine Tools." *Inventions* 3 (3): 41.

Cohen, J. "Breakthrough Technologies 2013." *MIT Technology Review. Retrieved October* 7: 2015.

Dong, S., P. Wang, and K. Abbas. 2021. "A Survey on Deep Learning and Its Applications." *Computer Science Review* 40. https://doi.org/10.1016/j.cosrev.2021.100379.

Duan, Lixiang, Mengyun Xie, Jinjiang Wang, and Tangbo Bai. 2018. "Deep Learning Enabled Intelligent Fault Diagnosis: Overview and Applications." *Journal of Intelligent & Fuzzy Systems* 35 (5): 5771–5784.

Ellefsen, Andre Listou, Vilmar Æsøy, Sergey Ushakov, and Houxiang Zhang. 2019. "A Comprehensive Survey of Prognostics and Health Management Based on Deep Learning for Autonomous Ships." *IEEE Transactions on Reliability* 68 (2): 720–740.

Farabet, Clement, Camille Couprie, Laurent Najman, and Yann LeCun. 2012. "Learning Hierarchical Features for Scene Labeling." *IEEE Transactions on Pattern Analysis and Machine Intelligence* 35 (8): 1915–1929.

Feng, Qiang, Xiujie Zhao, Dongming Fan, Baoping Cai, Yiqi Liu, and Yi Ren. 2019. "Resilience Design Method Based on Meta-Structure: A Case Study of Offshore Wind Farm." *Reliability Engineering & System Safety* 186: 232–244.

Glorot, Xavier, Antoine Bordes, and Yoshua Bengio. 2011. "Deep sparse rectifier neural networks." Proceedings of the Fourteenth International Conference on Artificial Intelligence and Statistics.

Goodfellow, Ian, Yoshua Bengio, and Aaron Courville. 2016. *Deep Learning.* MIT press.

Hamadache, Moussa, Joon Ha Jung, Jungho Park, and Byeng D. Youn. 2019. "A Comprehensive Review of Artificial Intelligence-Based Approaches for Rolling Element Bearing PHM: Shallow and Deep Learning." *JMST Advances* 1: 125–151.

Hinton, Geoffrey E., and Ruslan R. Salakhutdinov. 2006. "Reducing the Dimensionality of Data with Neural Networks." *Science* 313 (5786): 504–507.

Hirschberg, Julia, and Christopher D. Manning. 2015. "Advances in Natural Language Processing." *Science* 349 (6245): 261–266.

Hoang, Duy-Tang, and Hee-Jun Kang. 2019. "A Survey on Deep Learning Based Bearing Fault Diagnosis." *Neurocomputing* 335: 327–335.

Khan, Samir, and Takehisa Yairi. 2018. "A Review on the Application of Deep Learning in System Health Management." *Mechanical Systems and Signal Processing* 107: 241–265.

Krizhevsky, Alex, Ilya Sutskever, and Geoffrey E Hinton. 2012. "Imagenet Classification With Deep Convolutional Neural Networks." *Advances in Neural Information Processing Systems* 25.

LeCun, Yann, Yoshua Bengio, and Geoffrey Hinton. 2015. "Deep Learning." *Nature* 521 (7553): 436–444.

Lee, Jay, Fangji Wu, Wenyu Zhao, Masoud Ghaffari, Linxia Liao, and David Siegel. 2014. "Prognostics and Health Management Design for Rotary Machinery Systems—Reviews, Methodology and Applications." *Mechanical Systems and Signal Processing* 42 (1): 314–334. https://doi.org/10.1016/j.ymssp.2013.06.004.

Lei, Yaguo, Bin Yang, Xinwei Jiang, Feng Jia, Naipeng Li, and Asoke K. Nandi. 2020. "Applications of Machine Learning to Machine Fault Diagnosis: A Review and Roadmap." *Mechanical Systems and Signal Processing* 138: 106587.

Li, Chuan, Jose Valente De Oliveira, Mariela Cerrada, Diego Cabrera, René Vinicio Sánchez, and Grover Zurita. 2018. "A Systematic Review of Fuzzy Formalisms for Bearing Fault Diagnosis." *IEEE Transactions on Fuzzy Systems* 27 (7): 1362–1382.

Li, Dong, Yiqi Liu, and Daoping Huang. 2020. "Development of Semi-Supervised Multiple-Output Soft-Sensors With Co-Training and Tri-Training MPLS and MRVM." *Chemometrics and Intelligent Laboratory Systems* 199: 103970.

Liu, Ruonan, Boyuan Yang, Enrico Zio, and Xuefeng Chen. 2018. "Artificial Intelligence for Fault Diagnosis of Rotating Machinery: A Review." *Mechanical Systems and Signal Processing* 108: 33–47.

Minaee, S., Y. Y. Boykov, F. Porikli, A. J. Plaza, N. Kehtarnavaz, and D. Terzopoulos. 2022. "Image Segmentation Using Deep Learning: A Survey." *IEEE Transactions on Pattern Analysis and Machine Intelligence* 44 (7): 3523–3542. https://doi.org/10.1109/tpami.2021.3059968.

Nasiri, Sara, Mohammad Reza Khosravani, and Kerstin Weinberg. 2017. "Fracture Mechanics and Mechanical Fault Detection by Artificial Intelligence Methods: A Review." *Engineering Failure Analysis* 81: 270–293.

Rosenblatt, Frank. 1958. "The Perceptron: a Probabilistic Model for Information Storage and Organization in the Brain." *Psychological Review* 65 (6): 386.

Rumelhart, David E., Geoffrey E. Hinton, and Ronald J. Williams. 1986. "Learning Representations by Back-Propagating Errors." *Nature* 323 (6088): 533–536.

Sharp, Michael, Ronay Ak, and Thomas Hedberg Jr. 2018. "A Survey of the Advancing Use and Development of Machine Learning in Smart Manufacturing." *Journal of Manufacturing Systems* 48: 170–179.

Srivastava, N., G. Hinton, A. Krizhevsky, I. Sutskever, and R. Salakhutdinov. 2014. "Dropout: A Simple Way to Prevent Neural Networks from Overfitting." *Journal of Machine Learning Research* 15: 1929–1958.

Stetco, Adrian, Fateme Dinmohammadi, Xingyu Zhao, Valentin Robu, David Flynn, Mike Barnes, John Keane, and Goran Nenadic. 2019. "Machine Learning Methods for Wind Turbine Condition Monitoring: A Review." *Renewable Energy* 133: 620–635.

Sun, Shiliang, Chen Luo, and Junyu Chen. 2017. "A Review of Natural Language Processing Techniques for Opinion Mining Systems." *Information Fusion* 36: 10–25.

Tian, Yin, Dingfei Guo, Kunting Zhang, Lihao Jia, Hong Qiao, and Haichuan Tang. 2018. "A Review of Fault Diagnosis for Traction Induction Motor." 2018 37th Chinese Control Conference (CCC).

Wang, Jinjiang, Yulin Ma, Laibin Zhang, Robert X. Gao, and Dazhong Wu. 2018. "Deep Learning for Smart Manufacturing: Methods and Applications." *Journal of Manufacturing Systems* 48: 144–156.

Young, Tom, Devamanyu Hazarika, Soujanya Poria, and Erik Cambria. 2018. "Recent Trends in Deep Learning Based Natural Language Processing." *IEEE Computational Intelligence Magazine* 13 (3): 55–75.

Zhang, Wan, Min-Ping Jia, Lin Zhu, and Xiao-An Yan. 2017. "Comprehensive Overview on Computational Intelligence Techniques for Machinery Condition Monitoring and Fault Diagnosis." *Chinese Journal of Mechanical Engineering* 30: 782–795.

Zhang, Shen, Shibo Zhang, Bingnan Wang, and Thomas G Habetler. 2020. "Deep Learning Algorithms for Bearing Fault Diagnostics—A Comprehensive Review." *IEEE Access* 8: 29857–29881.

Zhao, Zhibin, Jingyao Wu, Tianfu Li, Chuang Sun, Ruqiang Yan, and Xuefeng Chen. 2021. "Challenges and Opportunities of AI-Enabled Monitoring, Diagnosis & Prognosis: A Review." *Chinese Journal of Mechanical Engineering* 34 (1): 56. https://doi.org/10.1186/s10033-021-00570-7.

Zhao, Rui, Ruqiang Yan, Zhenghua Chen, Kezhi Mao, Peng Wang, and Robert X. Gao. 2019. "Deep Learning and Its Applications to Machine Health Monitoring." *Mechanical Systems and Signal Processing* 115: 213–237.

Zhao, Guangquan, Guohui Zhang, Qiangqiang Ge, and Xiaoyong Liu. 2016. "Research advances in fault diagnosis and prognostic based on deep learning." 2016 Prognostics and System Health Management Conference (PHM-Chengdu).

Zou, Hui, and Trevor Hastie. 2005. "Regularization and variable Selection via the Elastic Net." *Journal of the Royal Statistical Society Series B: Statistical Methodology* 67 (2): 301–320.

I

Basic Applications of Deep Learning-Enabled Intelligent Fault Diagnosis

CHAPTER 2

Autoencoders for Intelligent Fault Diagnosis

INTRODUCTION

Induction motor, as one of the industrial power driving sources, occupies an important position in national economy and has been widely used to drive many kinds of machinery and industrial equipment, such as lifting hoist equipment, mining equipment, machine tools, etc. In order to guarantee normal operation of induction motors with timely maintenance, and avoid unnecessary loss, fault diagnosis on them is necessary (Benbouzid 2000; Huang and Yu 2013). However, due to environmental interference and inherent motor structure complexity, effective fault diagnosis for induction motors is challenging. Till date, various sensing techniques (Filippetti et al. 1998; Rodriguez, Belahcen, and Arkkio 2006; Pires et al. 2013) have been employed to measure certain physical quantities, such as current, vibration, radial and axial flux, rotor speed, etc., for the purpose of identifying induction motor faults. This is based on the fact that properties of induction motor may change when damages occur in it and such property change could be reflected in the measurement data. By extracting features from these measurement data, a classifier can then be built to distinguish different faults in induction motors. Therefore, the fault diagnosis problem can be converted into a classification problem which can be solved by machine-learning-based algorithms such as neural network (Malhi, Yan, and Gao 2011; Ahmadizar et al. 2015) and support vector machine (Li et al. 2013; Bordoloi and Tiwari 2014).

In the field of fault diagnosis, majority of machine learning algorithms are supervised learning algorithms which need a large amount of labeled data for training. Since neural networks possess strong representation ability due to their stacked hidden layers, they have been widely used as classifiers for machine fault diagnosis (Kuo 1995; Alexandru 2003; Su, Chong, and Kumar 2011). However, the training of neural networks also needs large amount of high-quality labeled data, and if the training samples are limited or cannot cover the testing distribution, the neural network may be easily overfitted which leads to a poor generalization especially for a complex classification problem. The induction motor

DOI: 10.1201/9781003474463-3

fault diagnosis belongs to such a challenging problem since the system itself is a complex electromechanical one. Therefore, some advanced signal processing technologies have been proposed for motor current or vibration signal analysis (Zarei and Poshtan 2007; Blödt et al. 2008; Puche-Panadero et al. 2009) so as to extract useful fault features for diagnosis. However, these state-of-the-art methods usually require the researchers to have a deep understanding of the induction motor system and the fault signals. In addition, the required expert knowledge is not easily obtained so that these methods are not general to address this task, which means these methods are not intelligent enough relative to the machine learning. In contrast, a deep "autoencoder" (AE) network structure has been trained to learn low-dimensional codes from high-dimensional input vectors in an unsupervised manner (Krizhevsky, Sutskever, and Hinton 2012). Furthermore, sparse AE (SAE) (Ranzato et al. 2007; Wright et al. 2010; Boureau et al. 2010) has been widely studied to realize deep learning, as it is highly effective for finding succinct and high-level representations of complex data.

Inspired by the prior research, this chapter presents an SAE-based deep neural network (DNN) approach for induction motor fault diagnosis. The rest of this chapter is organized as follows: AE and its variants, the details of experiments performed on a machine fault simulator and the corresponding results, and some conclusions drawn from this study.

AUTOENCODER AND ITS VARIANTS

This section is going to provide systematic introduction for AE, sparse AE, and denoising AE.

Autoencoder

An AE is a symmetrical neural network that can learn the features in an unsupervised manner by minimizing reconstruction errors (Shin et al. 2013). The basic structure of an AE is shown in Figure 2.1, where it tries to learn an approximation in the hidden layer so that the input data can be perfectly reconstructed at the output layer. However, the intrinsic problems of AE, such as simply copying input layer to hidden layer, make it ineffective in extracting meaningful features even though its output can be a perfect recovery of the input data.

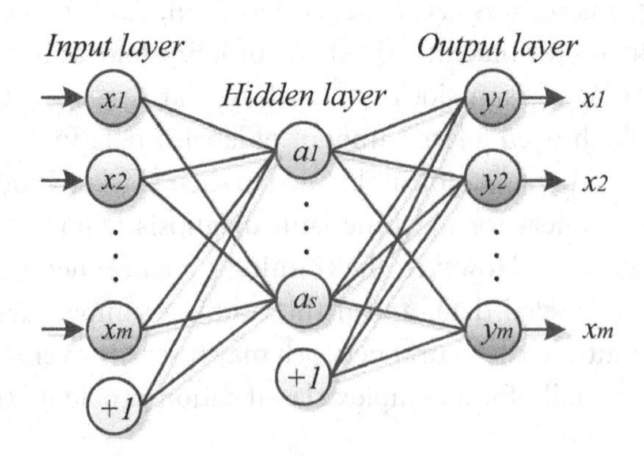

FIGURE 2.1 The structure of an autoencoder neural network.

Sparse Autoencoder

SAE, an extension of the AE, can learn relatively sparse features by introducing a sparse penalty term inspired by the sparse coding (Olshausen and Field 1996) into the AE. It can improve the performance of traditional AE and exhibits more practical application values.

From the measured vibration signals, an $N \times M$ data set can be constructed as $X=\{x(1),x(2),\ldots,x(i),\ldots,x(N)\},x(i)\in R^M$, where N is the number of data samples for each working condition and M is the length of each data sample. This data set is used as the input matrix of the SAE. Then a three-layer neural network similar to that shown in Figure 2.1 can be constructed, where the sigmoid function is chosen as activation function to the network. For the unlabeled input data matrix X, the goal is to learn and obtain a feature expression $h(x(i),W,b)=\sigma(Wx(i)+b),i=1,\ldots,N$ at the hidden layer so that the output $\sigma(W^T h(x(i),W,b)+c)$ is close to or refactoring the input. In the meantime, the sparse penalty term is added to the objective function of the AE so that the learned features are of the constraint rather than simply repeating input. The sparse penalty term actually works on the hidden layer to control the number of "active" neurons. In practice, if the output of a neuron is close to 1, the neuron is considered to be "active", otherwise it is considered "inactive". It is better to keep the neurons of the hidden layer "inactive" most of the time. Suppose that $a_j(X)$ denotes the activation of hidden unit j. In the forward propagation process, for a given input X, the activation of the hidden layer can be denoted as $a=sigmoid(WX+b)$, where W denotes the weights between the input layer and the hidden layer and b denotes the biases. Then the average activation of the hidden unit j can be given as:

$$\rho_j=\frac{1}{n}\sum_{i=1}^{n}\left[a_j(x(i))\right] \tag{2.1}$$

In this process, the average activation of each hidden neuron ρ_j is expected to be close to zero, namely the neurons of the hidden layer are mostly "inactive". To achieve this, the sparse term is added to the objective function that penalizes ρ_j if it deviates significantly from ρ. The penalty term is expressed as:

$$P_{penalty}=\sum_{j=1}^{S_2} K_L(\rho||\rho_j) \tag{2.2}$$

where S_2 is the number of neurons in the hidden layer. $KL(\cdot)$ is the Kullback–Leibler divergence (KL divergence) (Kullback and Leibler 1951), which can be written as:

$$KL(\rho||\rho_j)=\rho\log\frac{\rho}{\rho_j}+(1-\rho)\log\frac{1-\rho}{1-\rho_j} \tag{2.3}$$

This penalty function possesses the property that $KL(\rho||\rho_j)=0$ if $\rho=\rho_j$. Otherwise, it increases monotonically as ρ_j diverges from ρ, which acts as the sparsity constraint.

The cost function of the neural network is defined as:

$$C(W,b)=\left[\frac{1}{n}\sum_{i=1}^{n}\left(\frac{1}{2}\left\|h_{w,b}(x(i))-y(i)\right\|^{2}\right)\right]+\frac{\gamma}{2}\sum_{l=1}^{m_{l}-1}\sum_{i=1}^{s_i}\sum_{j=1}^{s_{i+1}}(W_{ij}(i)) \tag{2.4}$$

Adding the sparse penalty term to the cost function, it can be modified as:

$$C_{sparse}(W,b)=C(W,b)+\beta\sum_{j=1}^{S2}K_{L}(\rho\|\rho_{j}) \tag{2.5}$$

where β is the weight of the sparsity penalty.

During the coding process, the optimal parameters of W and b need to be identified. Since the sparse cost function shown in Eq. (2.5) is directly related to the parameters W and b, it can be solved by minimizing $C_{sparse}(W,b)$ to obtain these two parameters. This can be realized using the backpropagation algorithm (Rumelhart, Hinton, and Williams 1986), where the stochastic gradient descent approach is used for training and the parameters W and b in each iteration can be updated as:

$$W_{ij}(l)=W_{ij}(l)-\varepsilon\frac{\partial}{\partial W_{ij}(l)}C_{sparse}(W,b) \tag{2.6}$$

$$b_{i}(l)=b_{i}(l)-\varepsilon\frac{\partial}{\partial b_{i}(l)}C_{sparse}(W,b) \tag{2.7}$$

where ε is the learning rate. A forward pass on all training examples is used to compute the average activation ρ_j to get the sparse error, then the backpropagation algorithm works to update the parameters. After that, the effective sparse feature representations can be learned by the SAE.

Denoising Autoencoder

Vincent et al. (2008) proposed a denoising AE learning algorithm based on the idea of making the learned feature representation robust by adding partial corruption to the input pattern, which can be used to train stacked AEs to initialize the deep architectures. It has been known that this type of learning can help to obtain better feature representation with denoise coding, which motivates the introduction of denoising module into our study.

For the input matrix X, the process of denoising AE is shown in Figure 2.2, where the initial input X is corrupted to get a partially destroyed version χ by applying a stochastic mapping $\chi \sim qD(\chi|X)$, then χ is coded to y. Here distribution $qD(\cdot)$ can be described as follows: for each input $x(i)$, a fixed number of components are chosen randomly and their values are set to 0, while the others are kept unchanged. Then a joint distribution is defined as:

$$q^{0}(X,\chi,y)=q^{0}(X)qD(\chi|X)\delta_{f_{\theta}(X)}(y) \tag{2.8}$$

where $\delta_{f_{\theta}(X)}(y)$ is set to 0 if $f_{\theta}(x)\neq y$. $q^{0}(X)$ denotes the empirical distribution associated with N training inputs. Thus, y is a deterministic function of χ. Let θ be one parameter of

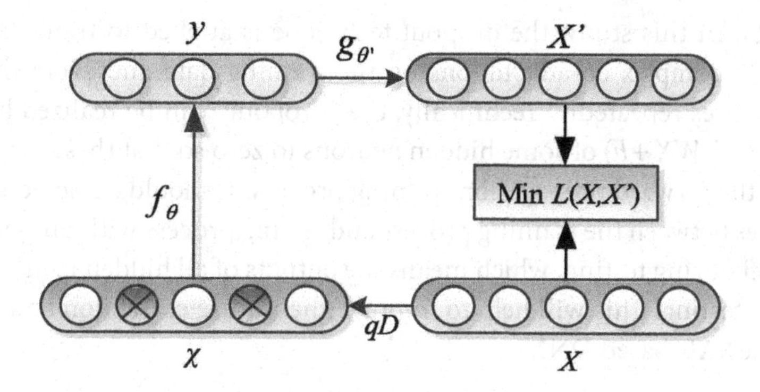

FIGURE 2.2 Structure of denoising autoencoder.

the joint distribution $q^0(X,\chi,y)$, it can be considered as a link parameter between χ and y. To obtain a reconstructed version (denoted as X') of the input matrix X from y, an objective function can be set as:

$$\arg\min_{\theta,\theta'} E_{q^0(X,X')}\left[L(X,X')\right] \tag{2.9}$$

where $L(X,X')$ is the reconstruction error. The reconstructed matrix X' can be obtained by minimizing function (2.9) using a stochastic gradient descent approach.

Specifically, to implement the denoising coding, some elements of the input X are randomly set to zero for generating the corrupted input χ. Then it is mapped with the basic AE to a hidden representation y, where $y = f_o(X) = sigmoid(WX+b)$ with $\theta = \{W,b\}$, from which X' can be reconstructed as $X' = g_{V'}(y) = Sigmoid(W_1 y + b_1)$. The reconstruction error $L(X,X')$ is computed from the difference between X' and X, and can be minimized by solving the cost function (2.9).

It should be noted that the motivation behind the corruption of the values of some components in X shares the same idea as of "salt noise" in image processing. Because unsupervised initialization of layers with an explicit denoising criterion can help to capture meaningful structure in the input distribution, the learned intermediate representations can contribute a lot to the subsequent learning tasks such as classification and clustering. Furthermore, it was indicated that denoising AE can improve the robustness of features and the sparsity of weights, thus making the discovery of interesting and prominent pattern easier (Cao, Huang, and Sun 2016).

SAE-BASED INTELLIGENT FAULT DIAGNOSIS

This section is going to introduce dropout and the proposed SAE-based fault diagnosis framework.

Dropout

Dropout is a technique that can help to reduce "overfitting" when training a neural network with limited training data set (Hinton et al. 2012). Generally, when the known training data set is small, the problem of overfitting occurs which leads to a poor performance on the test data. In this study, the dropout technique is applied to train the SAE-based DNN to prevent complex co-adaptations on the training data and avoid the extraction of the same features repeatedly. Technically, the "dropout" can be realized by setting the output $a = \text{sigmoid}(WX + b)$ of some hidden neurons to zero so that these neurons will not be involved in the forward propagation training process. It should be noted that there are some differences between the training process and testing process with dropout. The dropout is turned off during testing, which means the outputs of all hidden neurons will not be masked during testing. This will help to improve the feature extraction and classification capability of the SAE-based DNN.

SAE-Based Intelligent Fault Diagnosis

By taking advantage of the unsupervised learning, the denoising SAE is used to learn features from unlabeled data and initialize the DNN structure. Then the learned sparse feature representation of the AE is utilized to train a neural network classifier with a dropout module for induction motor fault diagnosis. Instead of direct utilization of learning representation by denoising SAE, the parameters of the neural network classifier with the corresponding parameters are initialized in a well-trained SAE and then updated further. This adaptation enables further fine-tuning of learned parameters so that the learned representation can capture more discriminative components in raw vibration signals. In addition, a previous study (Coates, Lee, and Ng 2011) indicated that the number of hidden neurons in the deep model is as important as the choice of learning algorithm and the depth of the model for achieving high classification performance, which means that single-layer network in unsupervised feature learning can also perform well. Therefore, in this study only single-layer SAE is utilized to realize the DNN for the induction motor fault diagnosis.

The training procedure of the DNN is shown in Figure 2.3 and it is described in subsequent text.

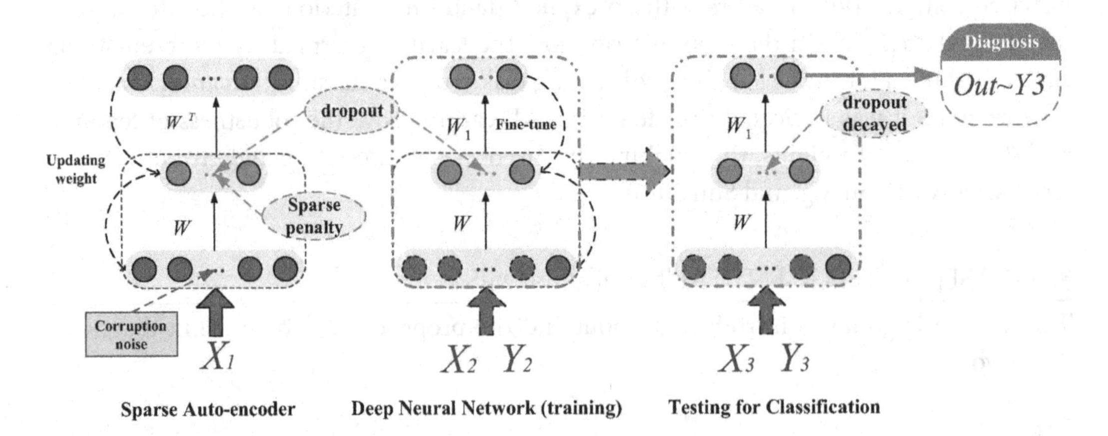

FIGURE 2.3 Structure of the sparse AE-based deep neural network.

The unlabeled induction motor vibration data $X1$ are first used to train the SAE though following steps:

1. Set up the learning rate, sparse rate and denoising parameters, dropout rate, etc. and initialize the weight W and b randomly.
2. Use the stochastic batches training method in the forward propagation algorithm to compute the average activation ρ_j for sparsity.
3. Compute the sparse cost function as:

$$C_{sparse}(W,b)=\left[\frac{1}{n}\sum_{i=1}^{n}\left(\frac{1}{2}\left\|h_{w,b}(x(i))-y(i)\right\|^2\right)\right]+\beta\sum_{j=1}^{52}KL(\rho\|\rho_j) \tag{2.10}$$

4. Update the parameters W and b based on Eq. (2.6) and (2.7).

Then the labeled induction motor vibration data ($X2$, $Y2$) are used to train the DNN for classification.

1. Use the parameters of the SAE to initialize the first layer of the DNN.
2. Set up the training parameters and dropout rate, and conduct the forward propagation algorithm to extract the labeled features for classification.
3. Compute the mean square error for the cost function of the DNN using Eq. (2.4).
4. Conduct the backpropagation algorithm the same as before except for the sparse term (set sparse penalty term to 0) to update the weights and fine-tune the entire network.

Finally, the test data set ($X3$, $Y3$) are used to verify the effectiveness of the presented SAE-based DNN.

EXPERIMENTAL STUDIES (FOR INDUCTION MOTOR INTELLIGENT FAULT DIAGNOSIS)

In this section, we are going to verify the performance of the proposed SAE-based intelligent fault diagnosis method on an induction motor dataset.

Experimental Description

To verify the effectiveness of the SAE-based DNN for induction motor fault diagnosis, the experiments have been conducted on a machine fault simulator as illustrated in Figure 2.4. The vibration signals can be acquired by an acceleration sensor when the motor operates under six different conditions. The vibration data of the simulator under the power supply frequency of 50 Hz are collected with the sampling frequency of 20 kHz. The descriptions of different motor working conditions are listed in Table 2.1.

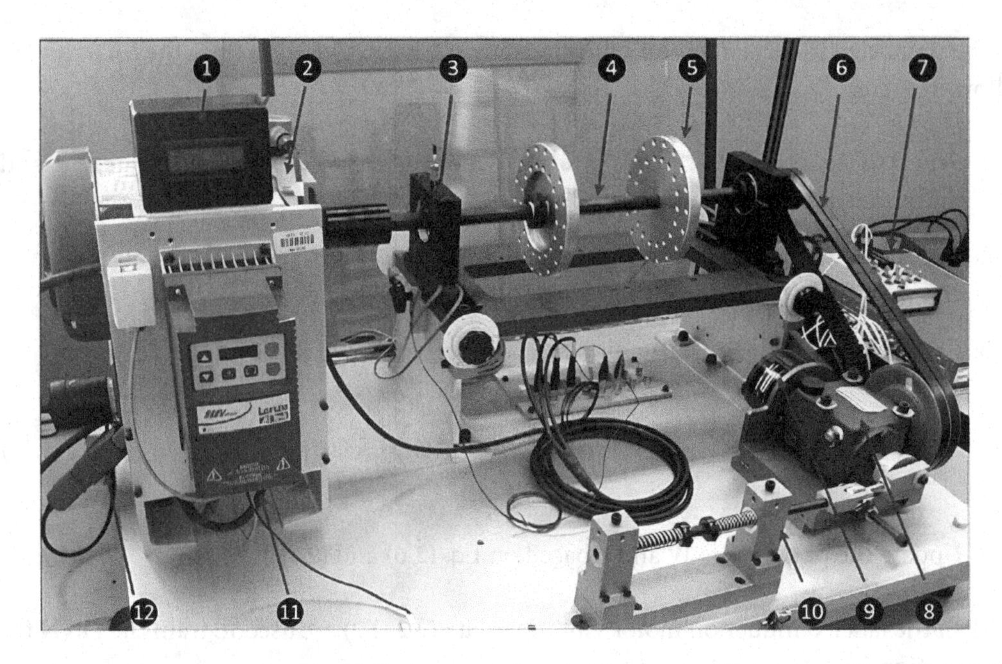

FIGURE 2.4 Experimental setup: (1) Tachometer, (2) induction motor, (3) bearing, (4) shaft, (5) load disc, (6) belt, (7) data acquisition board, (8) bevel gearbox, (9) magnetic load, (10) reciprocating mechanism, (11) variable speed controller, and (12) current probe.

The task defined here is to classify different motor working conditions based on vibration data. In this study, 600 data samples with 2000 features from each induction motor working condition were obtained, in which each feature corresponds to one sampled data point of the vibration. For each working condition, 400 of the samples were chosen randomly for training and the rest for testing. When 2400 training samples and 1200 testing samples are obtained with feature dimensionality of 2000, they are all normalized to set the value between −1 and 1.

Several neural network-based methods are compared here with the proposed approach. First, to verify the effectiveness of the SAE, normal neural networks without SAE are considered and three different model structures are considered. Two of them contain one hidden layer, which is the same as the proposed approach. Their hidden nodes are set as 100 and 600, respectively. They are denoted as NN(100) and NN(600). In addition, the neural network with two hidden layers, whose hidden nodes are 600 and 100, respectively, and denoted as NN(600–100), is also considered here. Then, two classifiers built on top of the

TABLE 2.1 Motor Condition Descriptions

	Condition	Description
HEA	Normal motor	Healthy, no defect
BRB	Broken bar	Three broken rotor bars
BRM	Bowed rotor	Rotor bent in center 0.01″
RMAM	Defective bearing	Inner race defect bearing in the shaft end
SSTM	Stator winding defect	3 turns shorted in stator winding
UBM	Unbalanced rotor	Unbalance caused by 3 added washers on the rotor

TABLE 2.2 Parameter Setting up of SAE-Based DNN

SAE			**DNN**		
M	2000	Input nodes	*M*	2000	Input nodes
S_2	600	Hidden nodes	*S*	600	Hidden nodes
Out	2000	Output nodes	*Out*	6	Output nodes
ρ	0.08	Sparse target	*Dropout*	0.3	Dropout nodes
β	0.4	Sparse rate	ε	1	Learning rate
Denoise	0.1	Denoising rate			
Dropout	0.3	Dropout rate			
ε	1	Learning rate			

SAE are also considered, i.e., support vector machine (SVM) and logistic regression (LR) (Khashei, Hamadani, and Bijari 2012). The features learned by the SAE, i.e., the hidden output of the SAE, are fed into both the SVM and the LR classifiers. For a fair comparison, the SAE is the same as the one used in the proposed approach, but the features learned by the SAE cannot be fine-tuned during training for both the SVM and the LR classifiers.

In the proposed SAE-based DNN framework, the input size, hidden node, and output size are set to 2000, 600, and 2000, respectively. Considering the vibration signals represent six different working conditions, the layer sizes for DNN is set as [2000 600 6]. The hidden nodes of the DNN and SAE are same so that the features learned by the SAE can be utilized by the DNN and fine-tuned further during supervised training. The three hyper-parameters of the proposed SAE-based DNN, including sparse parameter, denoted as sparse rate b, denoising rate, and dropout rate are set to be 0.4, 0.1, and 0.3, respectively, via cross-validation for comparison with other baseline methods. All parameters with their values of the proposed SAE-based DNN are listed in Table 2.2 for better understanding.

Here for all neural network-based methods, the cross-entropy error is calculated based on training samples and backpropagated through layers to update model parameters, in which the stochastic gradient descent method is used. The simulation environment for all algorithms performed in the experiments is MATLAB R2011b software environment.

Results and Discussion

Since the level of corruption noise is a hyper-parameter in the SAE, it is meaningful to study the effect of corruption noise strength on the final classification performance. Here, the noise levels changed from 0 to 0.5 with a step size of 0.1. Figure 2.5 (a) shows the classification performance with different corruption noise. It can be seen that proper denoising coding in the SAE could improve its performance, but too heavy corruption will degrade the input data quality, leading to a decrease of the classification performance. This verified that the SAE trained with appropriate noisy data can extract more robust features than traditional one and it is of great significance in practical complex environment.

In addition, the effect of dropout on performance of the DNN is also studied. Here, dropout rate changed from 0 to 0.5 with a step size of 0.1. Figure 2.5 (b) shows the classification performances with different dropout rates. The results showed that the best classification performance was obtained at a dropout rate close to 0.3 and the performance decreased

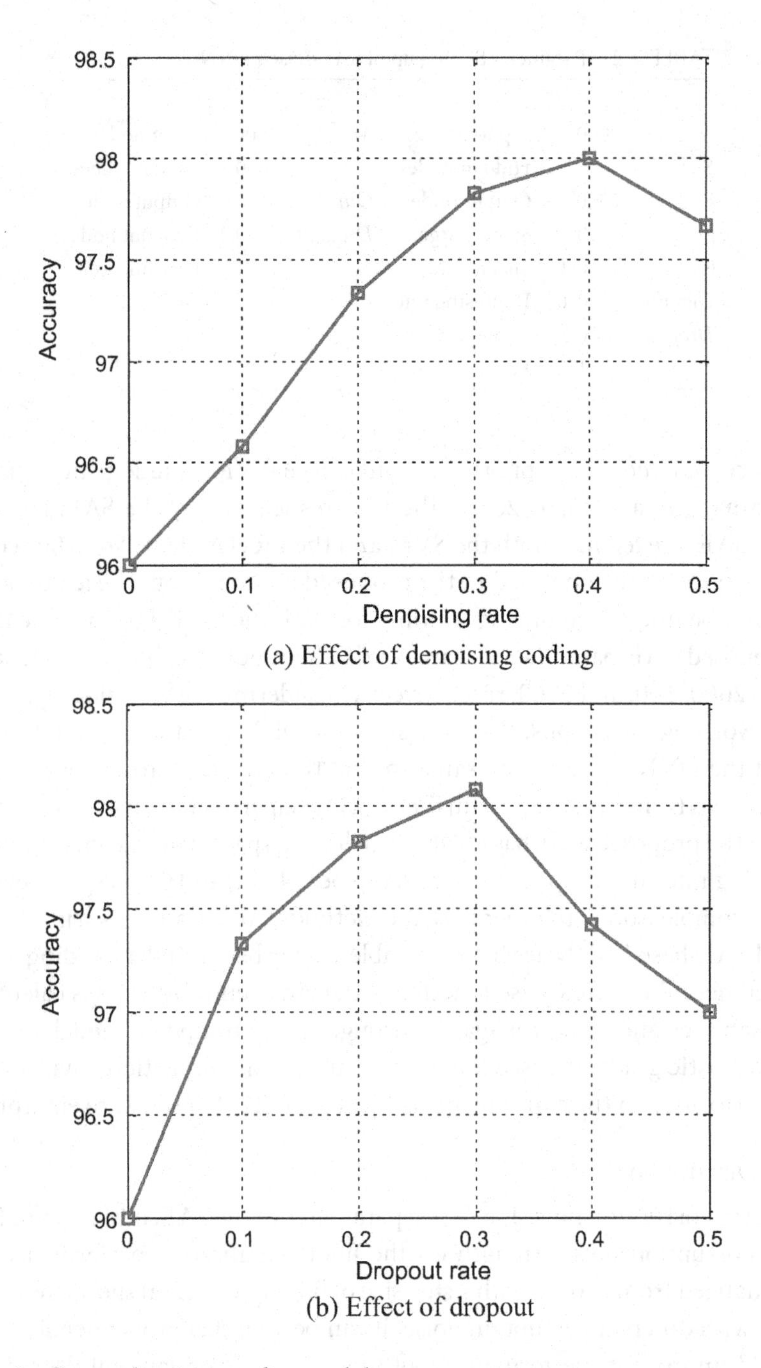

FIGURE 2.5 Studies on denoising coding and dropout.

when the dropout rate was more than 0.3. This indicated that dropout can improve the performance of the DNN, but too much dropout may lead to loss of some important neurons for feature representation. Based on results shown in Figure 2.5, it can be concluded that the SAE-based DNN approach presented in this study can realize the induction motor fault diagnosis and have a good classification accuracy of about 96% at least.

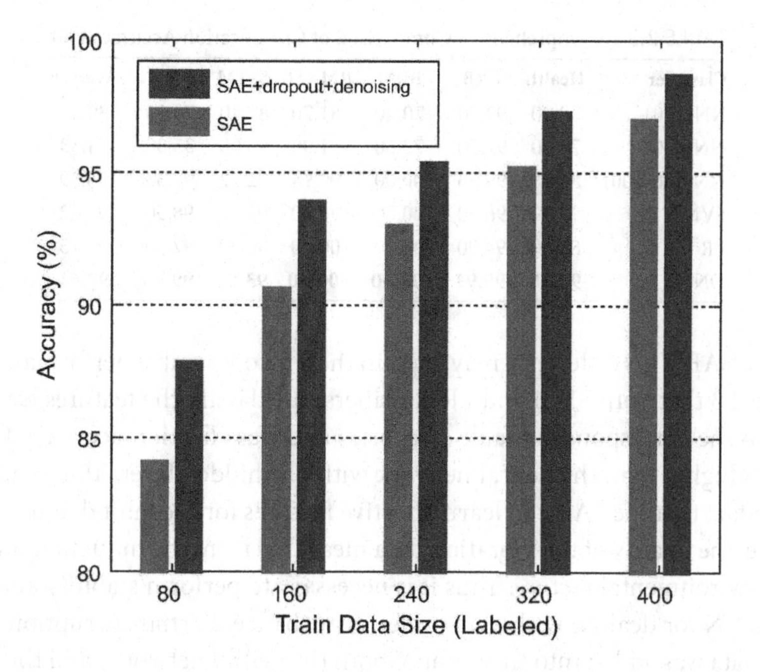

FIGURE 2.6 Comparison of two methods.

To verify the performance of presented approach that combines the sparse coding with denoising coding in the AE to learn features for the DNN and utilizes the "dropout" technique to overcome overfitting during the training process, a comparison study was then carried out, where traditional SAE was used as the baseline. The classification performance under different labeled training data set sizes were investigated, where the size of training data was changed from 80 to 400 with a step size of 80. The results are shown in Figure 2.6.

It can be seen that the presented approach has shown better performance than the SAE alone. In addition, the performance improvement achieved by the proposed approach for a small training data size is more significant than the one for a large training size, which verifies the robustness of the proposed approach. It may be explained that corruptions of input samples can be regarded as expanding data samples artificially. Furthermore, dropout can improve the performance of the DNN by preventing complex co-adaptations between the hidden neurons, especially for small training data size.

Comprehensive comparison results of the classification performance are shown in Table 2.3. It can be observed that the proposed SAE-based DNN framework achieves the best performance in all scenarios.

For the three neural network models without SAE, the best one is neural network with two hidden layers. This observation can be explained by the fact that the increased number of layers contributes to the learning capability of the neural network. However, NN(600–100) still performs worse than the proposed approach with one hidden layer. This result verifies the effectiveness of the SAE. The robust feature learned by the SAE can contribute to performance improvement of the following DNN.

Then, the proposed approach is also compared with two classifiers built on the top of the SAE. Different from the proposed approach, SVM and LR cannot fine-tune the features

TABLE 2.3 Comprehensive Comparison of Classification Accuracy (%)

Classifier	Health	BRB	BRM	RMAM	SSTM	UBM	Average
NN(100)	78.00	92.50	70.40	81.70	87.40	70.25	80.04
NN(600)	74.00	97.50	73.50	71.00	95.00	35.00	74.33
NN(600-100)	88.78	99.56	100.00	99.94	92.72	97.33	96.39
SVM	92.50	96.50	100.00	100.00	91.00	98.50	96.42
LR	83.50	94.00	100.00	100.00	81.50	97.50	92.75
DNN	**92.68**	**99.91**	**100.00**	**100.00**	**93.50**	**99.55**	**97.61**

learned by the SAE. This difference may explain their incomparable performance. It can be seen in Table 2.3 that both SVM and LR classifiers can classify the features learned by the SAE well into the corresponding fault categories and classification accuracy for either of them is much higher than the neural network with one hidden layer. This result is consistent with the fact that the SAE can learn effective features for the fault diagnosis.

In practice, the quality of the vibration data measured from the induction motor will be affected by environmental factors. Thus it is necessary to perform stability analysis of the SAE-based DNN for dealing with data under disturbance. Partial corruption at different levels to the data was added into the input of both the neural networks and the SAE-based DNN, and their influence on the classification accuracy is shown in Figure 2.7. When the degree of the data corruption increased from 0% to 40%, even though the performance of the SAE-based DNN decreases, the overall classification accuracy is still greater than 90%. In comparison, the performance of the neural networks decreases significantly, i.e., from 96.58% to 76.58% for the neural network with two hidden layers, and from 73.83% to 52.25% for the neural network with one hidden layer, reflecting that the SAE-based DNN possesses good stability against disturbance for induction motor fault diagnosis.

To demonstrate that the proposed approach is able to learn effective and discriminative representations for vibration signals automatically, the raw features, the features learned by the SAE and the features fine-tuned by the DNN are all visualized via a technique

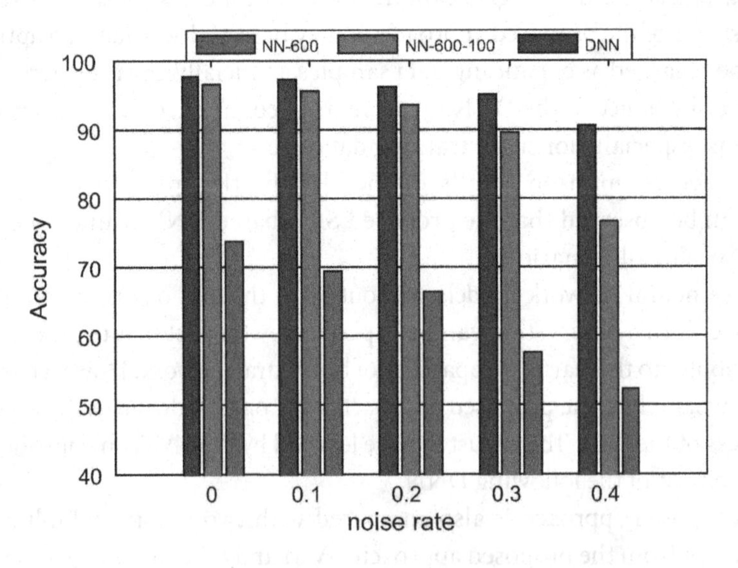

FIGURE 2.7 Histogram of classification accuracy with "noise" in input data.

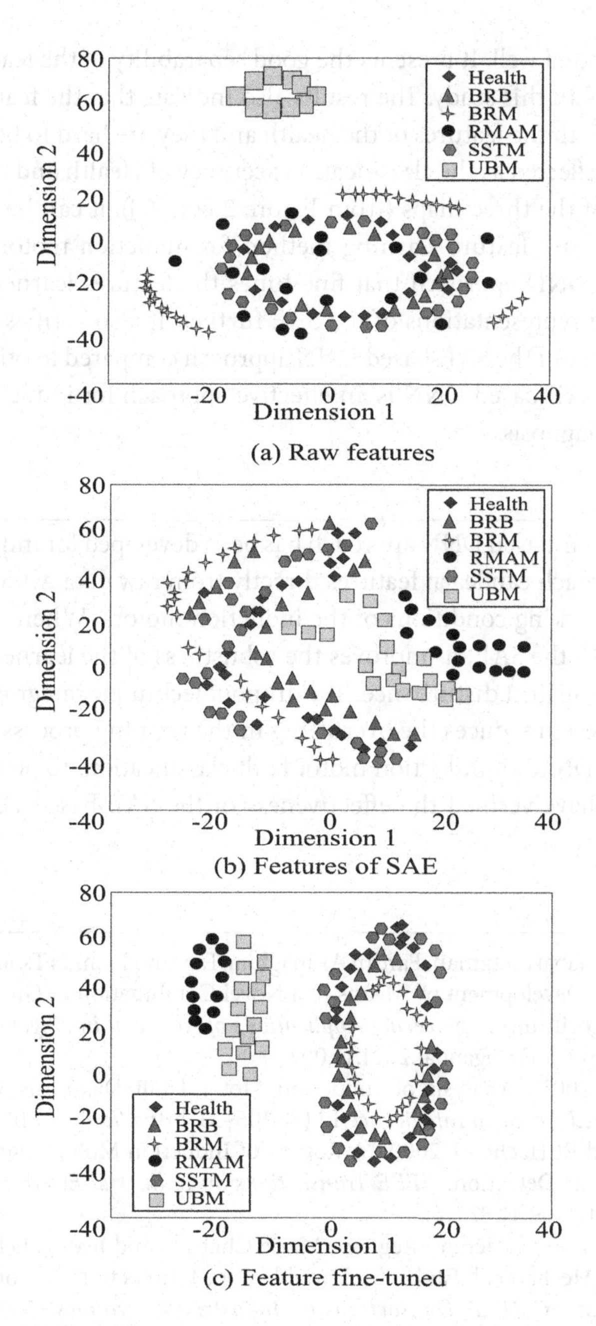

FIGURE 2.8 Feature visualization maps.

"t-SNE" (Van Der Maaten 2014), which is an effective data visualization technique for high-dimensional data. Here, principal component analysis (PCA) is used to reduce the dimensionality of the feature data to 50. This can speed up the computation of pairwise distances between the data points and suppresses some noise without severely distorting the interpoint distances. Then the dimensionality reduction technique "t-SNE" is used to convert the 50-dimensional representation to a 2-dimensional map and the resulting maps as a scatterplot have been shown in Figure 2.8 (a)–(c), respectively. It can be seen that the SAE features fine-tuned by the DNN cluster the best where data points of different conditions are separated well. It presents the good separability of the features extracted by the SAE-based DNN in this study. The results also indicate that the features of the SSTM is the most similar with the features of the health and they are hard to be separated well as others. This is also reflected as the classification accuracy of Health and SSTM in Table 2.3.

From the trend of the three maps (from Figure 2.8(a)–(c)), it can be observed that the SAE itself is an effective feature learning method for induction motor vibration signals and the SAE-based DNN approach that fine-tunes the features learned by the SAE can improve the learned representations of the SAE further. It also verifies the superior classification performance of the SAE-based DNN approach compared to other baseline methods. Therefore, the SAE-based DNN is an effective approach for induction motor feature learning and fault diagnosis.

CONCLUSION

In this chapter, an SAE-based DNN approach has been developed for induction motor fault diagnosis. This approach can learn features directly from raw data, which are discriminative over different working conditions of the induction motors. When the denoising coding is integrated with the SAE, it improves the robustness of the learned features and the stability of the DNN against disturbance. The dropout technique integrated into the neural network classifier design reduces the overfitting in the training process and improves the performance of the DNN for induction motor fault classification. Experimental study and a detailed analysis have verified the effectiveness of the SAE-based DNN for induction motor fault diagnosis.

REFERENCES

Ahmadizar, Fardin, Khabat Soltanian, Fardin AkhlaghianTab, and Ioannis Tsoulos. 2015. "Artificial Neural Network Development by Means of a Novel Combination of Grammatical Evolution and Genetic Algorithm." *Engineering Applications of Artificial Intelligence* 39 (March): 1–13. https://doi.org/10.1016/j.engappai.2014.11.003.

Alexandru, Monica. 2003. "Analysis of Induction Motor Fault Diagnosis With Fuzzy Neural Network." *Applied Artificial Intelligence* 17 (2): 105–33. https://doi.org/10.1080/713827102.

Benbouzid, Mohamed EI Hachemi. 2000. "A Review of Induction Motors Signature Analysis as a Medium for Faults Detection." *IEEE Transactions on Industrial Electronics* 47 (5): 984–993. https://doi.org/10.1109/41.873206.

Blödt, Martin, David Bonacci, Jeremi Regnier, Marie Chabert, and Jean Faucher. 2008. "On-Line Monitoring of Mechanical Faults in Variable-Speed Induction Motor Drives Using the Wigner Distribution." *IEEE Transactions on Industrial Electronics* 55: 522–533. https://doi.org/10.1109/TIE.2007.911941.

Bordoloi, D. J., and Rajiv Tiwari. 2014. "Support Vector Machine Based Optimization of Multi-Fault Classification of Gears with Evolutionary Algorithms from Time–Frequency Vibration Data." *Measurement* 55 (September): 1–14. https://doi.org/10.1016/j.measurement.2014.04.024.

Boureau, Y-Lan, Francis Bach, Yann LeCun, and Jean Ponce. 2010. "Learning Mid-Level Features for Recognition." 2010 IEEE Computer Society Conference on Computer Vision and Pattern Recognition. https://doi.org/10.1109/cvpr.2010.5539963.

Cao, Le-le, Wen-bing Huang, and Fu-chun Sun. 2016. "Building Feature Space of Extreme Learning Machine With Sparse Denoising Stacked-Autoencoder." *Neurocomputing* 174 (January): 60–71. https://doi.org/10.1016/j.neucom.2015.02.096.

Coates, Adam, Andrew Y. Ng, and Honglak Lee. 2011. "An Analysis of Single-Layer Networks in Unsupervised Feature Learning." International Conference on Artificial Intelligence and Statistics, International Conference on Artificial Intelligence and Statistics, June.

Filippetti, Fiorenzo, Giovanni Franceschini, Carla Tassoni, and Peter Vas. 1998. "AI Techniques in Induction Machines Diagnosis Including the Speed Ripple Effect." *IEEE Transactions on Industry Applications* 34 (1): 98–108. https://doi.org/10.1109/28.658729.

Hinton, Geoffrey E., Nitish Srivastava, Alex Krizhevsky, Ilya Sutskever, and Ruslan R. Salakhutdinov. 2012. "Improving Neural Networks by Preventing Co-Adaptation of Feature Detectors." *ArXiv Preprint ArXiv:1207.0580.*

Huang, Sunan, and Haoyong Yu. 2013. "Intelligent Fault Monitoring and Diagnosis in Electrical Machines." *Measurement* 46 (9): 3640–46. https://doi.org/10.1016/j.measurement.2013.07.004.

Khashei, Mehdi, Ali Zeinal Hamadani, and Mehdi Bijari. 2012. "A Novel Hybrid Classification Model of Artificial Neural Networks and Multiple Linear Regression Models." *Expert Systems With Applications* 39 (3): 2606–20. https://doi.org/10.1016/j.eswa.2011.08.116.

Krizhevsky, Alex, Ilya Sutskever, and Geoffrey E Hinton. 2012. "Imagenet Classification With Deep Convolutional Neural Networks." *Advances in Neural Information Processing Systems* 25.

Kullback, Solomon, and Richaed A. Leibler. 1951. "On Information and Sufficiency." *The Annals of Mathematical Statistics* 22: 79–86. https://doi.org/10.1214/aoms/1177729694.

Kuo, Ren Jieh. 1995. "Intelligent Diagnosis for Turbine Blade Faults Using Artificial Neural Networks and Fuzzy Logic." *Engineering Applications of Artificial Intelligence* 8 (1): 25–34. https://doi.org/10.1016/0952-1976(94)00082-X.

Li, Xu, A'nan Zheng, Xunan Zhang, Chenchen Li, and Li Zhang. 2013. "Rolling Element Bearing Fault Detection Using Support Vector Machine With Improved Ant Colony Optimization." *Measurement* 46 (8): 2726–34. https://doi.org/10.1016/j.measurement.2013.04.081.

Malhi, Arnaz, Ruqiang Yan, and Robert X. Gao. 2011. "Prognosis of Defect Propagation Based on Recurrent Neural Networks." *IEEE Transactions on Instrumentation and Measurement* 60: 703–11. https://doi.org/10.1109/tim.2010.2078296.

Olshausen, Bruno A., and David J. Field. 1996. "Emergence of Simple-Cell Receptive Field Properties by Learning a Sparse Code for Natural Images." *Nature* 381: 607–9. https://doi.org/10.1038/381607a0.

Pires, V. Fernão, Manuel Kadivonga, J. F. Martins, and A. J. Pires. 2013. "Motor Square Current Signature Analysis for Induction Motor Rotor Diagnosis." *Measurement* 46 (2): 942–48. https://doi.org/10.1016/j.measurement.2012.10.008.

Puche-Panadero, R., M. Pineda-Sanchez, M. Riera-Guasp, J. Roger-Folch, E. Hurtado-Perez, and J. Perez-Cruz. 2009. "Improved Resolution of the MCSA Method via Hilbert Transform, Enabling the Diagnosis of Rotor Asymmetries at Very Low Slip." *IEEE Transactions on Energy Conversion* 24 (1): 52–59. https://doi.org/10.1109/TEC.2008.2003207.

Ranzato, Marc'Aurelio, Y-Lan Boureau, and Yann Cun. 2007. "Sparse Feature Learning for Deep Belief Networks." Advances in Neural Information Processing Systems.

Rodriguez, P. V. Jover, Anouar Belahcen, and Antero Arkkio. 2006. "Signatures of Electrical Faults in the Force Distribution and Vibration Pattern of Induction Motors." *IEE Proceedings—Electric Power Applications* 153 (4): 523–529. https://doi.org/10.1049/ip-epa:20050253.

Rumelhart, David E., Geoffrey E. Hinton, and Ronald J. Williams. 1986. "Learning Representations by Back-Propagating Errors." *Nature* 323 (6088): 533–36. https://doi.org/10.1038/323533a0.

Shin, Hoo-Chang, M. R. Orton, D. J. Collins, S. J. Doran, and M. O. Leach. 2013. "Stacked Autoencoders for Unsupervised Feature Learning and Multiple Organ Detection in a Pilot Study Using 4D Patient Data." *IEEE Transactions on Pattern Analysis and Machine Intelligence* 35: 1930–43. https://doi.org/10.1109/tpami.2012.277.

Su, Hua, Kil T. Chong, and R. Ravi Kumar. 2011. "Vibration Signal Analysis for Electrical Fault Detection of Induction Machine Using Neural Networks." *Neural Computing and Applications* 20: 183–94. https://doi.org/10.1007/s00521-010-0512-3.

Van Der Maaten, Laurens. 2014. "Accelerating T-SNE Using Tree-Based Algorithms." *The Journal of Machine Learning Research* 15 (1): 3221–45.

Vincent, Pascal, Hugo Larochelle, Yoshua Bengio, and Pierre-Antoine Manzagol. 2008. "Extracting and Composing Robust Features with Denoising Autoencoders." *Proceedings of the 25th International Conference on Machine Learning - ICML '08*, 1096–1103. Helsinki, Finland: ACM Press. https://doi.org/10.1145/1390156.1390294.

Wright, John, Yi Ma, Julien Mairal, Guillermo Sapiro, Thomas S. Huang, and Shuicheng Yan. 2010. "Sparse Representation for Computer Vision and Pattern Recognition." *Proceedings of the IEEE* 98: 1031–44. https://doi.org/10.1109/jproc.2010.2044470.

Zarei, Jafar, and Javad Poshtan. 2007. "Bearing Fault Detection Using Wavelet Packet Transform of Induction Motor Stator Current." *Tribology International* 40 (5): 763–69. https://doi.org/10.1016/j.triboint.2006.07.002.

CHAPTER 3

Deep Belief Networks for Intelligent Fault Diagnosis

INTRODUCTION

Failures often occur in manufacturing machines which may cause disastrous accidents such as economic losses, environmental pollution, and even casualties. Effective diagnosis of these failures is essential in order to enhance reliability and reduce costs of operation and maintenance of the manufacturing equipment. As a result, research on fault diagnosis of manufacturing machines that utilizes data acquired by advanced sensors and makes decisions using processed sensor data has been successful in various applications (Gao et al. 2015; Riera-Guasp, Antonino-Daviu, and Capolino 2015; Chen et al. 2016). Induction motors, as the source of actuation, have been widely used in many manufacturing machines, and their working states directly influence system performance, thus affecting the production quality. Therefore, proper acquisition of data reflecting the working states of induction motors can help in early identification of potential failures (Drif and Marques Cardoso 2014). During recent years, various approaches for induction motor fault diagnosis have been developed and innovated continuously (Antonino-Daviu et al. 2013; Faiz, Ghorbanian, and Ebrahimi 2014; Karvelis et al. 2015; Wang et al. 2016).

Artificial intelligence (AI)-based fault diagnosis techniques have been widely studied, and have succeeded in many applications of electrical machines and drives (Matić et al. 2012; Zhang et al. 2016). For example, a two-stage learning method including sparse filtering and neural networks was proposed to develop an intelligent fault diagnosis method which could learn features from raw signals (Lei et al. 2016). The feed-forward neural network using Levenberg–Marquardt algorithm showed a new way to detect and diagnose induction machine faults (Boukra, Lebaroud, and Clerc 2013), where the results were not affected by the load condition and the fault types. In another study, a special structure of support vector machine (SVM) was proposed, which combined directed acyclic graph-SVM (DAG-SVM) with recursive undecimated wavelet packet transform, for inspection of broken rotor bar fault in induction motors (Keskes and Braham 2015). Fuzzy system and Bayesian theory

DOI: 10.1201/9781003474463-4

were utilized in machine health monitoring by C. Chen, Zhang, and Vachtsevanos (2012). Although these studies have shown the advantages of AI-based approaches for induction motor fault diagnosis, most of these approaches are based on supervised learning, in which high-quality training data with good coverage of true failure conditions are required to perform model training (Murphey et al. 2006). However, it is not easy to obtain sufficient labeled fault data to train the model in practice.

Furthermore, many fault diagnosis tasks in induction motors depend on feature extraction from the measured signals. The feature characteristics directly affect the effectiveness of fault recognition. In the existing literature, many feature extraction methods are suitable for fault diagnosis tasks, such as time-domain statistical analysis, frequency-domain spectral analysis (Wang, Gao, and Yan 2014), and time-scale/frequency analysis (Boashash 2015), among which wavelet analysis (Yan, Gao, and Chen 2014), which belongs to time-scale analysis, is a powerful tool for feature extraction and has been well applied to processing non-stationary signals. Whereas, the problem is that different features extracted from these methods may affect the classification accuracy. Therefore, an automatic and unsupervised feature learning from the measured signals for fault diagnosis is needed.

Limitations above can be overcome by deep learning algorithms which follow an effective way of learning multiple layers of representations (Hinton 2007b). Essentially, a deep learning algorithm that uses deep neural networks which contain multiple hidden layers to learn information from the input was developed, but was not put into practice because of its training difficulty until Geoffrey Hinton proposed layer-wise pre-training algorithm to effectively train deep networks in 2006 (Hinton and Salakhutdinov 2006). Since then, deep learning techniques have advanced significantly and their successful applications have been observed in various fields (Arel, Rose, and Karnowski 2010), including hand written digit recognition (Bengio 2009), computer vision (Jia et al. 2014; Szegedy et al. 2015; Cai et al. 2016; He et al. 2022), Google Map (Hinton 2007a), and speech recognition (Le 2013; Deng, Hinton, and Kingsbury 2013; LeCun, Bengio, and Hinton 2015). In addition, for natural language processing (NLP), deep learning has achieved several successful applications and made significant contributions to its progress (Deng and Yu 2014; Tai, Socher, and Manning 2015; Xiong, Merity, and Socher 2016). In the area of fault diagnosis, deep learning theory also has many applications. For example, deep neural network built for fault signature extraction was utilized for bearings and gearboxes (Jia et al. 2016), while a classification model based on deep network architecture was proposed in the task of characterizing health states of the aircraft engine and electric power transformer (Tamilselvan and Wang 2013). The deep belief network (DBN) was also used for identifying faults in reciprocating compressor valves (Tran, AlThobiani, and Ball 2014). Sparse coding was used to build deep architecture for structural health monitoring (Guo et al. 2014), and a unique automated fault detection method named "Tilear" using deep learning concepts was proposed for the quality inspection of electromotor (Sun et al. 2014). Furthermore, autoencoder-based DBN model was successfully applied to quality inspection (Sun, Steinecker, and Glocker 2014), while a sparse model based on autoencoder was shown to form a deep architecture, which realized induction motor fault diagnosis (Sun et al. 2016).

Inspired by the prior research, this chapter presents a deep learning model based on DBN for induction motor fault diagnosis. The deep model is built on restricted Boltzmann machine (RBM) which is the building unit of a DBN and by stacking multiple RBMs one by one, the whole deep network architecture can be constructed. It can learn high-level features from frequency distribution of measured signals for diagnosis tasks. Including this section, this chapter is organized in five sections. Section 3.2 provides theoretical background of the deep learning algorithm. Section 3.3 presents the proposed fault diagnosis approach, where the deep architecture based on DBN is described in detail. Experiments to verify the effectiveness of the proposed deep model are detailed in Section 3.4, where classification performance is discussed. Section 3.5 summarizes the whole study and gives future directions.

THEORETICAL FRAMEWORK

DBN is a deep architecture with multiple hidden layers that has the capability to learn hierarchical representations automatically in an unsupervised way and perform classification at the same time. In order to accurately structure the model, it contains both unsupervised pre-training procedure and supervised fine-tuning strategy. Generally, it is difficult to learn a large number of parameters in a deep architecture which has multiple hidden layers due to the vanishing gradient problem. To address this issue, an effective training algorithm, which learns one layer at a time and each pair of layers is seen as one RBM model, is proposed which was introduced by Mohamed, Yu, and Deng (2010) and Chen and Lin (2014). As DBN is formed by units of RBM, the basic unit of DBN, i.e., RBM, is introduced first.

Architecture of RBM

RBM is a commonly used mathematical model in probability statistics theory and follows the theory of log-linear Markov random field (MRF) (Tran, AlThobiani, and Ball 2014) which has several special forms and RBM is one of them. An RBM model contains two layers: One layer is the input layer which is also called visible layer, and the other layer is the output layer which also called hidden layer. RBM can be represented as a bipartite undirected graphical model. All the visible units of the RBM are fully connected to hidden units, while units within one layer do not have any connections between each other. That is to say, there are no connection between visible units or between hidden units. The architecture of a RBM is shown in Figure 3.1.

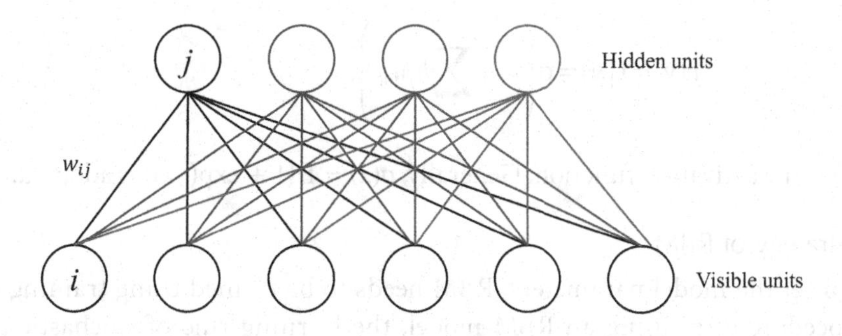

FIGURE 3.1 Architecture of restricted Boltzmann machine.

In Figure 3.1, v represents the visible layer, i is the i-th visible unit, h is the hidden layer, and j is the j-th hidden unit. Connections between these two layers are undirected. An energy function is proposed to describe the joint configuration (v, h) between them, which is expressed as:

$$E(v,h)=-\sum_{i\in visible} a_i v_i - \sum_{j\in hidden} b_j h_j - \sum_{i,j} v_i h_j w_{ij}. \quad (3.1)$$

Here, v_i, and h_j represent the visible unit i and hidden unit j, respectively; a_i and b_j are their biases. w_{ij} denotes the weight between these two units. Therefore, the joint distribution of this pair can be obtained using the energy function where θ is the model parameter set containing a, b, and w:

$$p(v,h)=\frac{1}{Z(\theta)}\exp\left(-E(v,h)\right), \quad (3.2)$$

$$Z(\theta)=\sum_v\sum_h \exp\left(-E(v,h)\right). \quad (3.3)$$

Due to the particular connections in RBM model, it satisfies conditional independence. Therefore, conditional probability of this pair of layers can be written as:

$$p(h|v)=\prod_i p(h_i|v), \quad (3.4)$$

$$p(v|h)=\prod_j p(v_i|h). \quad (3.5)$$

Mathematically,

$$p(h_j=1|v)=\sigma\left(b_j+\sum_i v_i w_{ij}\right), \quad (3.6)$$

$$p(v_i=1|h)=\sigma\left(a_i+\sum_j h_j w_{ij}\right), \quad (3.7)$$

where $\sigma(x)$ is the activation function. Generally, $\sigma(x) = 1/(1 + \exp(-x))$ is adopted.

Training Strategy of RBM

In order to set the model parameters, RBM needs to be trained using training dataset. In the procedure of training an RBM model, the learning rule of stochastic gradient descent is adopted. The log-likelihood probability of the training data is calculated, and its derivative with respect to the weights is seen as the gradient, shown in Eq. (3.8). The goal of this training procedure is to update network parameters in order to obtain a convergence model.

$$\frac{\partial \log p(v)}{\partial w_{ij}} = \langle v_i h_j \rangle_{data} - \langle v_i h_j \rangle_{\text{model}}. \tag{3.8}$$

Parameter update rules were originally derived by Hinton and Sejnowki, which can be written as:

$$\Delta_{w_{ij}} = \varepsilon\left(\langle v_i h_j \rangle_{data} - \langle v_i h_j \rangle_{\text{model}}\right), \tag{3.9}$$

where ε is the learning rate, the symbol $\langle \cdot \rangle_{data}$ represents an expectation from the data distribution while the symbol $\langle \cdot \rangle_{model}$ is an expectation from the distribution defined by the model. The former term is easy to compute exactly, while the latter one is intractable to compute (Salakhutdinov and Hinton 2009).

An approximation to the gradient is used to obtain the latter one which is realized by performing alternating Gibbs sampling, as illustrated in Figure 3.2(a).

Later, a fast-learning procedure was proposed, which starts with the visible units, then all the hidden units are computed at the same time using Eq. (3.6). After that, visible units are updated in parallel to get a "reconstruction" by Eq. (3.7), as illustrated in Figure 3.2(b), and the hidden units are updated again (Hinton 2012). Model parameters are updated as:

$$\Delta w_{ij} = \varepsilon\left(\langle v_i h_j \rangle_{data} - \langle v_i h_j \rangle_{recon}\right). \tag{3.10}$$

In addition, for practical problems that come down to real-valued data, Gaussian–Bernoulli RBM is introduced to deal with this issue. Input units of this model are linear while hidden units are still binary. Learning procedure for Gaussian–Bernoulli RBM is very similar to binary RBM introduced above.

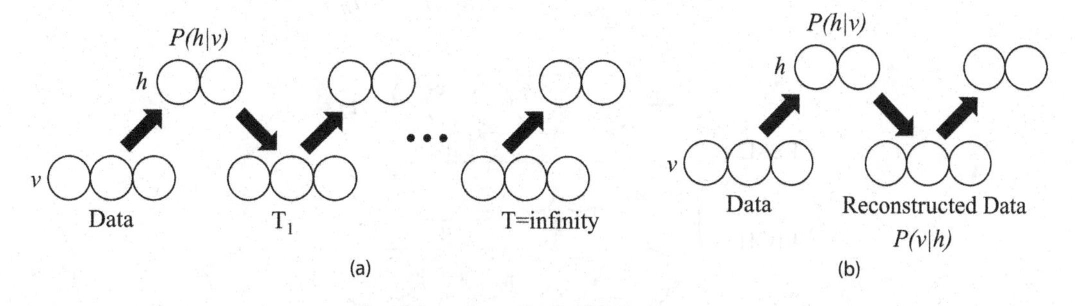

FIGURE 3.2 (a) Alternating Gibbs sampling and (b) a quick way to learn RBM.

DBN-BASED INTELLIGENT FAULT DIAGNOSIS

DBN Architecture

DBN model is a deep network architecture with multiple hidden layers which contain many nonlinear representations. It is a probabilistic generative model and can be formed by multiple RBMs as shown in Figure 3.3. It illustrates the way of stacking one RBM on top of another. DBN architecture can be built by stacking multiple RBMs one by one to form a deep network architecture.

As DBN has multiple hidden layers, it can learn from the input data and extract hierarchical representation corresponding to each hidden layer. Joint distribution between visible layer v and the l hidden layers h_k can be calculated mathematically from conditional distribution $P(h^{k-1}|h^k)$ for the $(k-1)$ th layer conditioned on the k-th layer and visible-hidden joint distribution $P(h^{n-1}, h^n)$:

$$P\left(v,h^1,\ldots,h^n\right)=\left(\prod_{k=1}^{n-1}P\left(h^{k-1}\middle|h^k\right)\right)P\left(h^{n-1},h^n\right). \tag{3.11}$$

For deep neural networks, learning such amount of parameters using traditional supervised training strategy is impractical because errors transferred to low level layers will faint through several hidden layers and the ability to adjust the parameters is weak for traditional backpropagation method. It is difficult for the network to generate globally optimal parameters. Here the greedy layer-by-layer unsupervised pre-training method is used for training DBNs. This procedure can be illustrated as follows: The first step is to train the input units (v) and the first hidden layer (h_1) using RBM rule (denoted as RBM1). Next, the first hidden layer (h_1) and the second hidden layer (h_2) are trained as an RBM (denoted as RBM2) where the output of RBM1 is used as the input for RBM2. Similarly, following hidden layers can be trained as RBM3, RBM4, …, RBMn until the set number of layers are

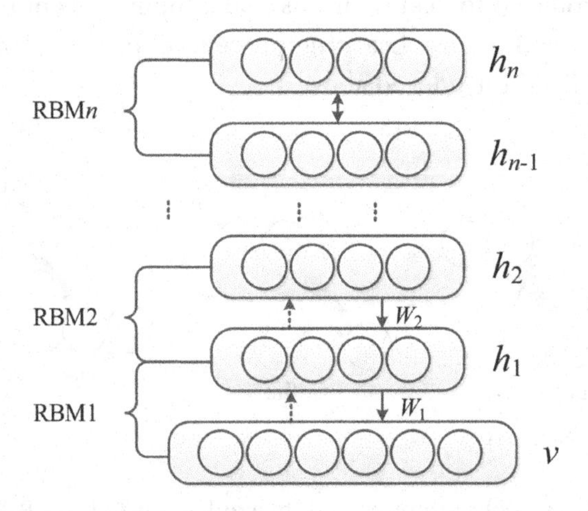

FIGURE 3.3 Architecture of deep belief network.

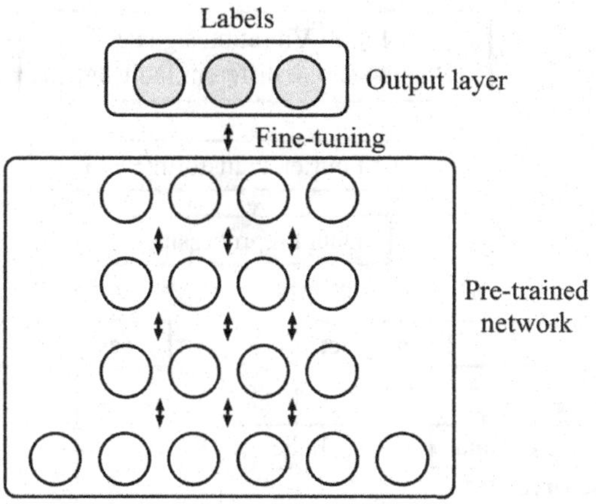

FIGURE 3.4 Supervised fine-tuning process.

met. It is an unsupervised pre-training procedure which gives the network an initialization that contributes to convergence on the global optimum.

For classification tasks, fine-tuning of all the parameters of this deep architecture together is needed after the layer-wise pre-training, as shown in Figure 3.4. It is a supervised learning process using labels to eliminate the training error and improve the classification accuracy (Bengio et al. 2006; Hinton, Osindero, and Teh 2006).

DBN-Based Intelligent Fault Diagnosis

Based on DBN, a fault diagnosis approach for induction motor has been developed, as illustrated in Figure 3.5, where the DBN model is built to extract multiple levels of representation from the training dataset.

Vibration signals are selected as the input of the whole system for fault diagnosis as they usually contain useful information that can reflect the working state of induction motors. However, there exists correlation between sampled data points. This is difficult for DBN architecture to model as it does not have the ability to function the correlation between the input units which may influence the following classification task. Therefore, in this study the vibration signals are transformed from time domain to frequency domain using fast Fourier transform (FFT), and then frequency distribution of each signal is used as the input of the DBN architecture. This is beneficial to classification task during the training procedure. Specifically, DBN learns a model that generates input data, which can obtain more intrinsic characteristics of the input, thus improving classification accuracy eventually. In this module, DBN stacked by a number of RBMs is built and then trained by training dataset from data preparation module to obtain the model parameters. The DBN training process is shown in Figure 3.6. Input parameters of the architecture will be first initialized including a set of neuron numbers and hidden layer numbers, together with training epochs. Each layer of the architecture is then trained as an RBM unit, and the output of lower-layer RBM is used as the training input for the next-layer RBM.

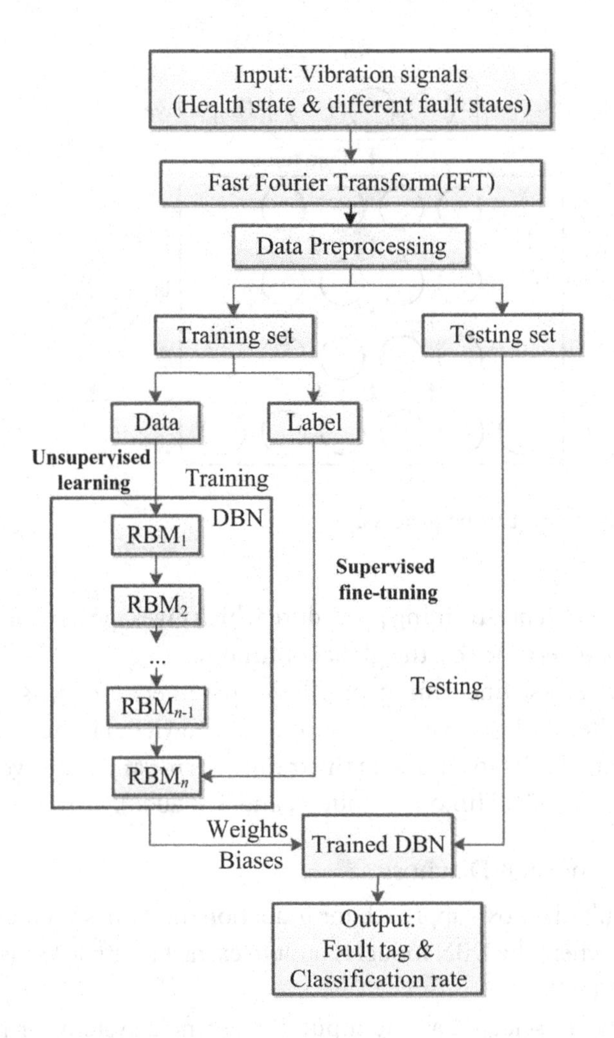

FIGURE 3.5 Deep belief network-based fault diagnosis procedure.

After layer-by-layer learning, synaptic weights and biases are settled and the basic structure is determined. Classification process is then followed to predict the fault category. It is a supervised fine-tuning procedure and the proposed method adopts the backpropagation training algorithm to realize fine-tuning which uses labeled data for training, so that it can improve the discriminative ability for classification task. The unsupervised training process trains one RBM at a time and afterwards supervised fine-tuning process using labels adjusts weights of the whole model. The difference between DBN outputs and the target label is regarded as training error. In order to obtain the minimum error, the deep network parameters will be updated based on learning rules.

After training the DBN model, all the DBN parameters are fixed, and the next procedure is to test the classification capability of the trained DBN model and classification rate is calculated as an index for evaluation. The vibration signal is the input of the constructed fault diagnosis system, and its output indicates working states of the induction motor.

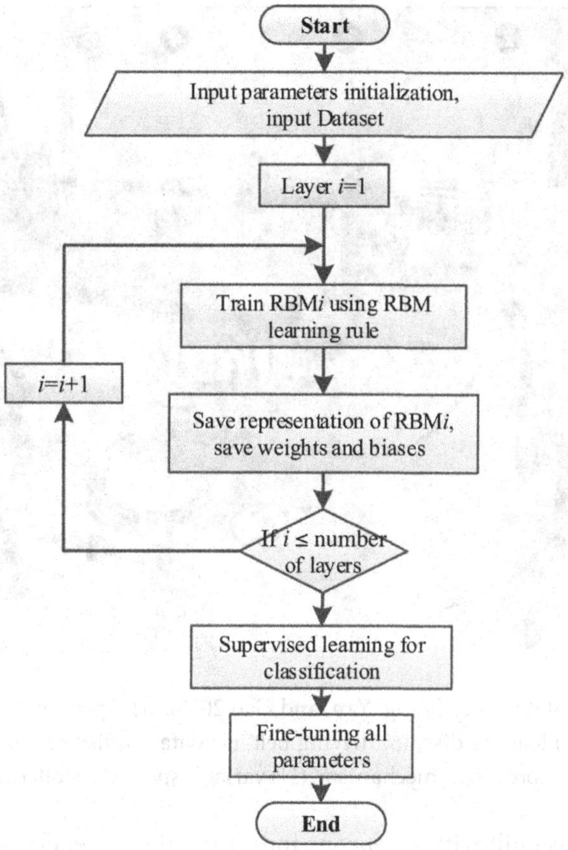

FIGURE 3.6 Training process of the deep belief network model.

EXPERIMENTAL STUDIES FOR INDUCTION MOTOR INTELLIGENT FAULT DIAGNOSIS

Experimental Description

Experimental Setting

To evaluate the proposed approach for fault diagnosis of induction motors, experimental studies are conducted using a machine fault simulator illustrated in Figure 3.7. It simulates six different conditions during motor operation and vibration signals are measured corresponding to different working states. The descriptions of each operation conditions are listed in Table 3.1 (Yang, Yan, and Gao 2015).

These acquired vibration signals are used to test the DBN-based fault diagnosis system. These vibration signals are divided into training datasets and testing datasets separately, and both datasets are randomized before being used in the DBN model.

Comparison Approaches

According to Hinton's theory (LeCun, Bengio, and Hinton 2015), parameters of the DBN architecture are initialized in advance. The input layer has 1000 units for vibration signals,

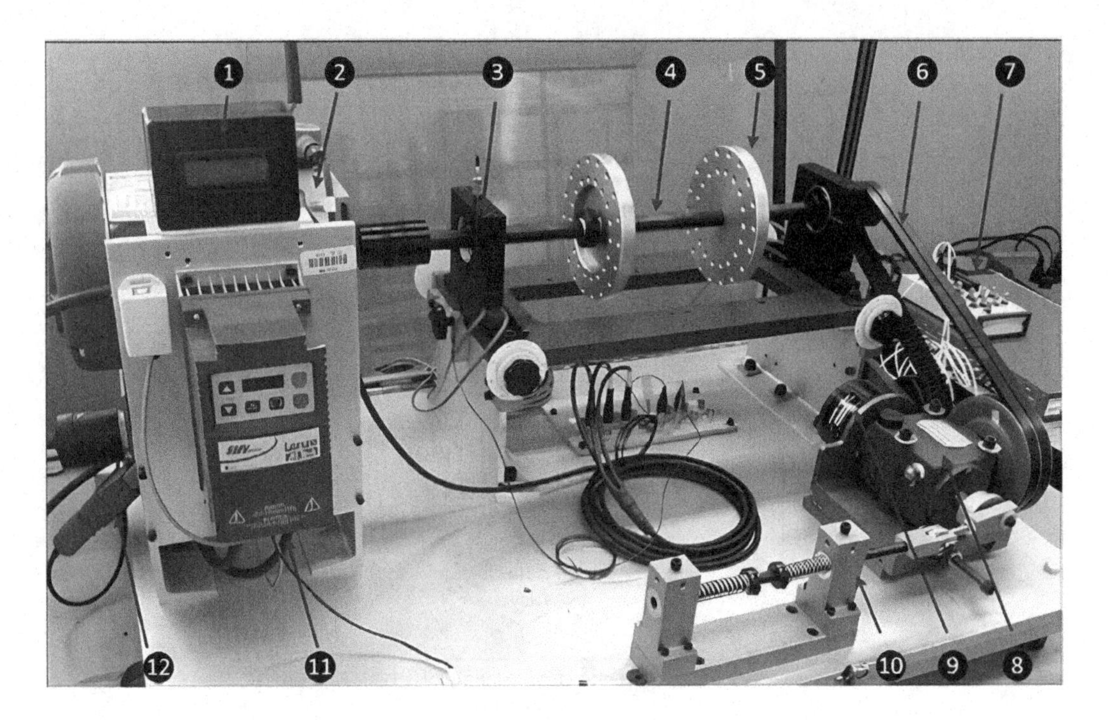

FIGURE 3.7 Experimental facility (Yang, Yan, and Gao 2015): (1) Opera meter, (2) induction motor, (3) bearing, (4) shaft, (5) loading disc, (6) driving belt, (7) data acquisition board, (8) bevel gearbox, (9) magnetic load, (10) reciprocating mechanism, (11) variable speed controller, and (12) current probe.

and the output layer is built with 6 neurons indicating the target classes corresponding to 6 different operation conditions. The deep model has 4 hidden layers with each containing 500 neurons. Training epochs is set to be 100, and learning rate of the RBM learning process and fine-tuning procedure are set as 0.01 and 0.1, respectively. Threshold value of training error is set to be 0.12.

For the number of hidden units in each layer, networks with small number of hidden units may not learn enough representations for future tasks while networks with large numbers of neurons may increase the possibility of overfitting which cause poor generalization in untrained dataset. From the literature, there is no formula to calculate an exact number of neurons being used, but the number of neurons within a range is effective in practice. As the input neurons are 1000, number of units for each hidden layer is selected as 500 to avoid both too narrow and too complicated network structures. In addition, the

TABLE 3.1 Motor Condition Descriptions (Yang, Yan, and Gao 2015)

Condition		Description
HEA	Normal motor	Healthy motor without defect
SSTM	Stator winding defect	Three turns shorted in stator winding
UBM	Unbalanced rotor	Unbalance caused by three added washers on the rotor
RMAM	Defective bearing	Inner race defect bearing in the shaft end
BRB	Broken bar	Broken rotor bars
BRM	Bowed rotor	Rotor bent in center 0.01″

relationship between numbers of hidden units and classification performance of the network is also discussed in the next section.

In order to verify the effectiveness of the proposed approach in actual applications of fault diagnosis for induction motors, comparative experiments have been carried out, some of which are listed here:

1. Original vibration signals are used directly as input of soft-max function;
2. Original vibration signals are used directly as input of the backpropagation network with one hidden layer;
3. Original vibration signals are preprocessed to extract time domain features including mean value, root mean square (RMS) value, shape factor, skewness, kurtosis, impulse factor, and crest factor (Riera-Guasp, Antonino-Daviu, and Capolino 2015), then seven selected features are used as input of the BP network;
4. Four features including shape factor, impulse factor, crest factor, and kurtosis are used as input of the BP network,
5. Signals are preprocessed with 5-layer wavelet packet decomposition to get 63 sub-frequency bands, then the energy features at all sub-frequency bands are used as input of the BP network.

In addition, another comparative experiment is carried out where unprocessed raw vibration signal is used directly as the input data.

Results and Discussion

Results of Datasets

In this validation experiment, training dataset and testing dataset contain vibration signals from all six working states. The proposed DBN-based fault diagnosis system is used to classify these six different working states at the same time. All learning algorithms are repeated 50 times and the average classification rates are calculated, as listed in Table 3.2. In this case, training dataset has 1200 samples (200 samples for each working state), while testing dataset has 600 samples (100 samples for each working state). Classification rate achieved in testing datasets is 99.98%.

TABLE 3.2 Classification Rate with Different Methods

Model	**Classification Rate (%)**
Soft-max	17.88
BP network	80.67
Seven-time domain features + BP	85.61
Four-time domain features +BP	81.39
Wavelet packet analysis + BP	95.33
DBN	95.67
FFT + DBN	99.98

The results from a comparative study are also listed in Table 3.2. As per the diagnosis results, the first method failed in the fault diagnosis task. Using original vibration signals without preprocessing, BP network with single hidden layer cannot achieve accurate classification. For time-domain analysis, different features used in the tests give different results, which means accurate classification needs manual intervention to pick proper features to do the task of fault diagnosis. The wavelet analysis method provided similar results as compared to the DBN using unprocessed vibration data, but it needs the signal preprocessing first and the results also rely on whether the extracted features are good for the task, while the DBN using frequency distribution of the signals achieved the highest classification rate in all experiments. In addition, the proposed DBN-based approach combines feature learning and classification together to improve the efficiency of fault diagnosis. These experiments proved that proposed approach is an effective way for fault diagnosis of induction motors.

For traditional fault diagnosis approaches, as the raw vibration signal always contains too much noise interference, one essential step is the data preprocessing to eliminate noise and extract the relevant information for classification. Hence, a robust and effective feature extraction requires some high-quality engineering experience and professional knowledge that is often challenging and hard to obtain. Compared with traditional fault diagnosis approaches, DBN-based deep learning architecture can automatically learn representations from the input and reduce the manual work so that it can reduce the influence of artificial factors.

Figure 3.8 shows detailed label distribution in the verification experiment using DBN model and FFT-DBN model, respectively. The results indicate that FFT-DBN model has better classification capability in the task of fault diagnosis for induction motors than DBN model as FFT-DBN model only has 1 misclassification sample while DBN model has confusion in label 2, 3, and 5. It also illustrates that frequency distribution of the signal is suitable in the application of DBN model, while DBN architecture cannot well model the temporal information of input data which may influence the following classification process.

In Figure 3.9, the training error and the classification rates of these two models are shown. From the comparison, FFT-DBN model has faster convergence and better classification rate. There is a fluctuation during the learning process in DBN model which means the architecture may not be stable enough to learn an accurate model for the classification task.

Effects of Scales and Depths of DBN Architecture

In this section, experiments are conducted to study the relationship between classification performance and different deep architectures in induction motor applications. Both DBN model using time-domain signals and FFT-DBN model using frequency distribution of the signals are investigated, and the comparison results are provided and discussed.

In the experiment, hidden neurons from 10 to 100 and from 100 to 1500 are considered. The hidden layers are explored as deep as six layers. Each group of experiments are repeated 50 times and the average classification rates are calculated as the evaluation index for deep architecture. Both DBN model and FFT-DBN model are tested, and the results are shown in Figure 3.10.

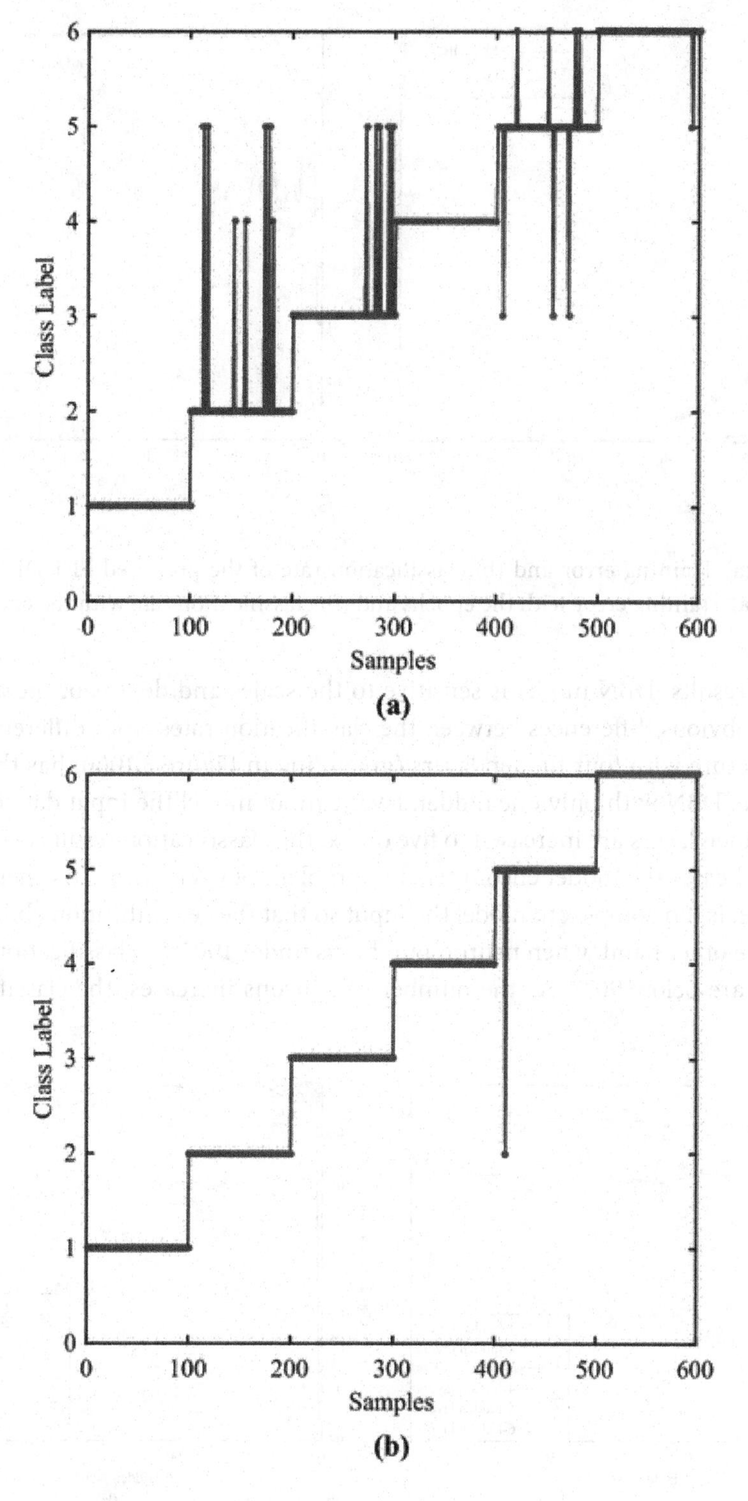

FIGURE 3.8 Label distribution of testing dataset. (a) DBN model using time-domain vibration signal. (b) FFT-DBN model using frequency-domain signal.

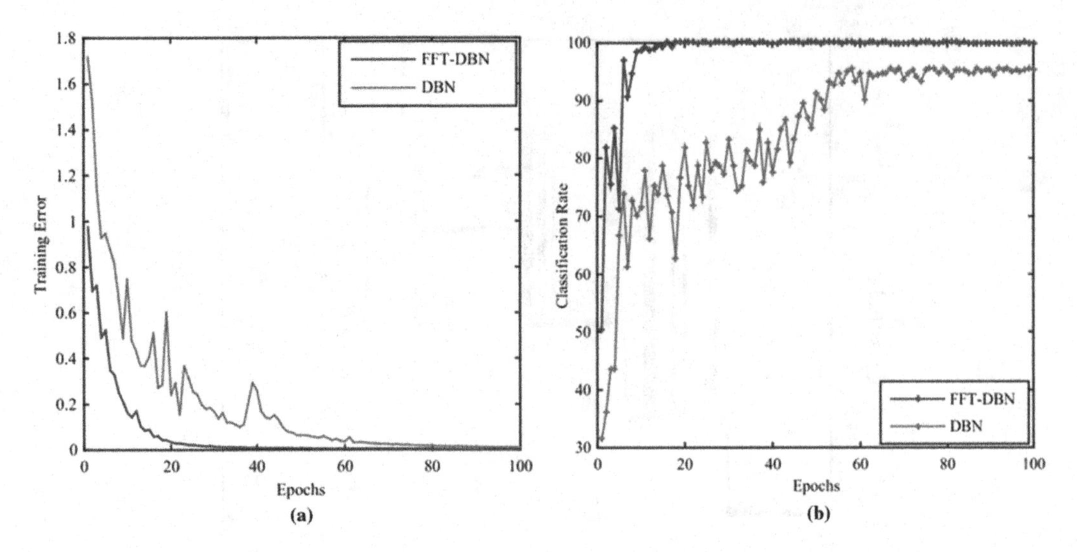

FIGURE 3.9 (a) Training error and (b) classification rate of the proposed FFT-DBN model and DBN model. (a) Training error with the epochs and (b) classification rate with the epochs.

From the results, DBN model is sensitive to the scales and depths of the architecture as there are obvious differences between the classification rates from different networks. DBN architecture with four hidden layers (green line in Figure 3.10(a)) has the best classification rate. DBN with only one hidden layer cannot model the input data exactly, and when the hidden layers are increased to five or six, the classification results become unstable which indicates the model encounters the problem of overfitting. In other words, the trained model is too complex to model the input so that the generalization ability becomes worse. On the other hand, when neuron number is under 100, the classification rates from DBN model are below 90%. As the number of neurons increases, the classification rate

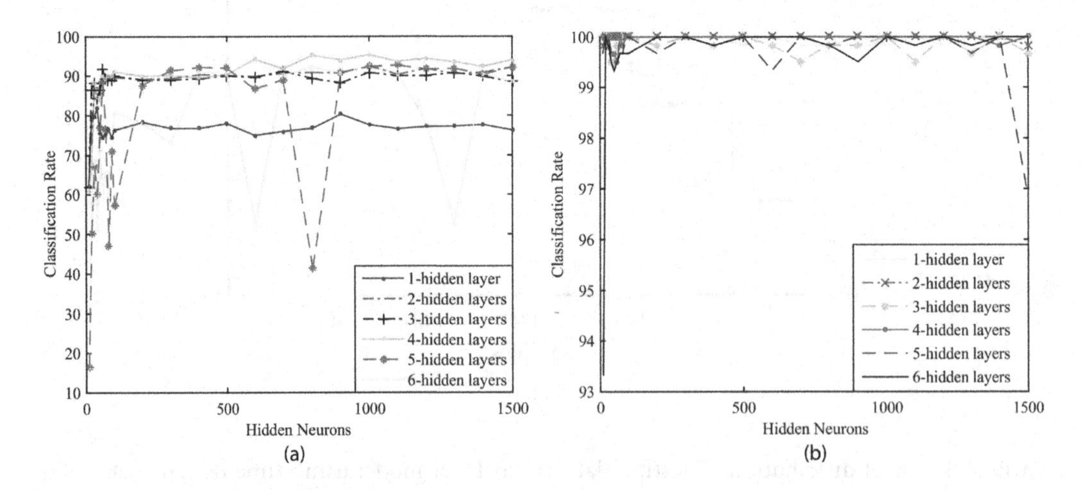

FIGURE 3.10 Different classification results with different hidden layers and hidden neurons both in (a) DBN model and (b) FFT-DBN model.

improves and when the number of neurons increases to 1000, the classification rate begins to decrease, indicating too much neurons may cause overfitting that influences classification capability of the model.

Compared with classification results of the DBN model, the results of the FFT-DBN model are much more stable, as shown in Figure 3.10(b). There is little fluctuation with the increase of hidden neurons, but the classification rates are all above 99% except two extreme individuals. One is a network with six hidden layers and each hidden layer has ten hidden units which is too narrow to learn enough representations, and the other is the network of 5 hidden layers with 1500 neurons at each hidden layer which has the possibility of overfitting as the input data is not so complex. However, generally, FFT-DBN model performances well in various network structures, both in accuracy and stability.

From the comparison, DBN model using time-domain signals has less classification rates in various architectures than the one using frequency distribution of the signals, which means DBN architecture cannot well model signals that correlate between input units. Lacking of time-domain information leads to an inaccurate model of the input data. Therefore, using frequency distribution as input to the DBN architecture gives a good alternative solution in fault diagnosis task for induction motors.

CONCLUSION

This chapter presents a deep learning model based on DBN, where frequency distribution of the measured data is used as input for fault diagnosis of induction motors in manufacturing. This deep architecture uses RBM as a building unit, and uses greedy layer-wise training for model construction. The presented approach makes use of strong capabilities of DBN, which can model high-dimensional data and learn multiple layers of representation, thus reducing training error and improving classification accuracy. Experimental studies are carried out using vibration signals to verify the effectiveness of the DBN model for feature learning, providing a new way of feature extraction for automatic fault diagnosis in manufacturing.

REFERENCES

Antonino-Daviu, Jose, Selin Aviyente, Elias G. Strangas, and Martin Riera-Guasp. 2013. "Scale Invariant Feature Extraction Algorithm for the Automatic Diagnosis of Rotor Asymmetries in Induction Motors." *IEEE Transactions on Industrial Informatics* 9 (1): 100–8. https://doi.org/10/f4jrs8.

Arel, Itamar, Derek C. Rose, and Thomas P. Karnowski. 2010. "Research Frontier: Deep Machine Learning—A New Frontier in Artificial Intelligence Research." *IEEE Computational Intelligence Magazine* 5 (4): 13–8. https://doi.org/10/fgfnv3.

Bengio, Yoshua. 2009. "Learning Deep Architectures for AI." *Foundations and Trends in Machine Learning* 2 (1): 1–127. https://doi.org/10/cxmtqb.

Bengio, Yoshua, Pascal Lamblin, Dan Popovici, and Hugo Larochelle. 2006. "Greedy Layer-Wise Training of Deep Networks." In *Advances in Neural Information Processing Systems*. Vol. 19. MIT Press. https://proceedings.neurips.cc/paper/2006/hash/5da713a690c067105aeb2fae32403405-Abstract.html.

Boashash, Boualem. 2015. *Time-Frequency Signal Analysis and Processing: A Comprehensive Reference*. Academic Press.

Boukra, Tahar, Abdesselam Lebaroud, and Guy Clerc. 2013. "Statistical and Neural Network Approaches for the Classification of Induction Machine Faults Using the Ambiguity Plane Representation." *IEEE Transactions on Industrial Electronics* 60 (9): 4034–42. https://doi.org/10/ggpj9w.

Cai, Yingfeng, Hai Wang, Xiaobo Chen, Li Gao, and Long Chen. 2016. "Vehicle Detection Based on Visual Saliency and Deep Sparse Convolution Hierarchical Model." *Chinese Journal of Mechanical Engineering* 29 (4): 765–72. https://doi.org/10/gstnpm.

Chen, Xue-Wen, and Xiaotong Lin. 2014. "Big Data Deep Learning: Challenges and Perspectives." *IEEE Access* 2: 514–25. https://doi.org/10/ghhfhh.

Chen, Juan, Jiming Ma, Jia Li, and Yongling Fu. 2016. "Performance Optimization of Grooved Slippers for Aero Hydraulic Pumps." *Chinese Journal of Aeronautics* 29 (3): 814–23. https://doi.org/10/gsgckq.

Chen, Chaochao, Bin Zhang, and George Vachtsevanos. 2012. "Prediction of Machine Health Condition Using Neuro-Fuzzy and Bayesian Algorithms." *IEEE Transactions on Instrumentation and Measurement* 61 (2): 297–306. https://doi.org/10/cfsqfv.

Deng, Li, Geoffrey Hinton, and Brian Kingsbury. 2013. "New Types of Deep Neural Network Learning for Speech Recognition and Related Applications: An Overview." 2013 IEEE International Conference on Acoustics, Speech and Signal Processing, 8599–8603. https://doi.org/10/gfrbhk.

Deng, Li, and Dong Yu. 2014. "Deep Learning: Methods and Applications." *Foundations and Trends in Signal Processing* 7 (3–4): 197–387. https://doi.org/10/gcz66f.

Drif, M'hamed, and Antonio J. Marques Cardoso. 2014. "Stator Fault Diagnostics in Squirrel Cage Three-Phase Induction Motor Drives Using the Instantaneous Active and Reactive Power Signature Analyses." *IEEE Transactions on Industrial Informatics* 10 (2): 1348–60. https://doi.org/10/gstnnd.

Faiz, Jawad, Vahid Ghorbanian, and Bashir Mahdi Ebrahimi. 2014. "EMD-Based Analysis of Industrial Induction Motors with Broken Rotor Bars for Identification of Operating Point at Different Supply Modes." *IEEE Transactions on Industrial Informatics* 10 (2): 957–66. https://doi.org/10/gstnn4.

Gao, Huizhong, Lin Liang, Xiaoguang Chen, and Guanghua Xu. 2015. "Feature Extraction and Recognition for Rolling Element Bearing Fault Utilizing Short-Time Fourier Transform and Non-Negative Matrix Factorization." *Chinese Journal of Mechanical Engineering* 28 (1): 96–105. https://doi.org/10/gstnm9.

Guo, Junqi, Xiaobo Xie, Rongfang Bie, and Limin Sun. 2014. "Structural Health Monitoring by Using a Sparse Coding-Based Deep Learning Algorithm With Wireless Sensor Networks." *Personal and Ubiquitous Computing* 18 (8): 1977–87. https://doi.org/10/f6p66s.

He, You, Hesheng Tang, Yan Ren, and Anil Kumar. 2022. "A Deep Multi-Signal Fusion Adversarial Model Based Transfer Learning and Residual Network for Axial Piston Pump Fault Diagnosis." *Measurement* 192 (March): 110889. https://doi.org/10/grtf4h.

Hinton, Geoffrey E. 2007a. "To Recognize Shapes, First Learn to Generate Images." *Progress in Brain Research* 165:535–47. https://doi.org/10.1016/S0079-6123(06)65034-6.

———. 2007b. "Learning Multiple Layers of Representation." *Trends in Cognitive Sciences* 11 (10): 428–34. https://doi.org/10/bwbtmp.

———. 2012. "A Practical Guide to Training Restricted Boltzmann Machines." In *Neural Networks: Tricks of the Trade: Second Edition*, edited by Montavon, Grégoire, Geneviève B. Orr, and Klaus-Robert Müller, 599–619. Lecture Notes in Computer Science. Berlin, Heidelberg: Springer. https://doi.org/10.1007/978-3-642-35289-8_32.

Hinton, Geoffrey E., Simon Osindero, and Yee-Whye Teh. 2006. "A Fast Learning Algorithm for Deep Belief Nets." *Neural Computation* 18 (7): 1527–54. https://doi.org/10/cjnhxz.

Hinton, G. E., and R. R. Salakhutdinov. 2006. "Reducing the Dimensionality of Data With Neural Networks." *Science* 313 (5786): 504–7. https://doi.org/10/b7vgmj.

Jia, Feng, Yaguo Lei, Jing Lin, Xin Zhou, and Na Lu. 2016. "Deep Neural Networks: A Promising Tool for Fault Characteristic Mining and Intelligent Diagnosis of Rotating Machinery With Massive Data." *Mechanical Systems and Signal Processing* 72–73 (May): 303–15. https://doi.org/10/gfx4xw.

Jia, Yangqing, Evan Shelhamer, Jeff Donahue, Sergey Karayev, Jonathan Long, Ross Girshick, Sergio Guadarrama, and Trevor Darrell. 2014. "Caffe: Convolutional Architecture for Fast Feature Embedding." Proceedings of the 22nd ACM International Conference on Multimedia, 675–78. MM '14. New York, USA: Association for Computing Machinery. https://doi.org/10/gcsgr3.

Karvelis, Petros, George Georgoulas, Ioannis P. Tsoumas, Jose Alfonso Antonino-Daviu, Vicente Climente-Alarcón, and Chrysostomos D. Stylios. 2015. "A Symbolic Representation Approach for the Diagnosis of Broken Rotor Bars in Induction Motors." *IEEE Transactions on Industrial Informatics* 11 (5): 1028–37. https://doi.org/10/f7szts.

Keskes, Hassen, and Ahmed Braham. 2015. "Recursive Undecimated Wavelet Packet Transform and DAG SVM for Induction Motor Diagnosis." *IEEE Transactions on Industrial Informatics* 11 (5): 1059–66. https://doi.org/10/f7s52x.

Le, Quoc V. 2013. "Building High-Level Features Using Large Scale Unsupervised Learning." 2013 IEEE International Conference on Acoustics, Speech and Signal Processing, 8595–98. https://doi.org/10/ggdt85.

LeCun, Yann, Yoshua Bengio, and Geoffrey Hinton. 2015. "Deep Learning." *Nature* 521 (7553): 436–44. https://doi.org/10/bmqp.

Lei, Yaguo, Feng Jia, Jing Lin, Saibo Xing, and Steven X. Ding. 2016. "An Intelligent Fault Diagnosis Method Using Unsupervised Feature Learning Towards Mechanical Big Data." *IEEE Transactions on Industrial Electronics* 63 (5): 3137–47. https://doi.org/10/gfx4tm.

Matić, Dragan, Filip Kulić, Manuel Pineda-Sánchez, and Ilija Kamenko. 2012. "Support Vector Machine Classifier for Diagnosis in Electrical Machines: Application to Broken Bar." *Expert Systems With Applications* 39 (10): 8681–89. https://doi.org/10/gstnn5.

Mohamed, Abdel-Rahman, Dong Yu, and L. Deng. 2010. "Investigation of Full-Sequence Training of Deep Belief Networks for Speech Recognition." In INTERSPEECH 2010, ISCA, 2846–49. Chiba, Japan. https://doi.org/10/gstnpx.

Murphey, Yi Lu, M. A. Masrur, ZhiHang Chen, and Baifang Zhang. 2006. "Model-Based Fault Diagnosis in Electric Drives Using Machine Learning." *IEEE/ASME Transactions on Mechatronics* 11 (3): 290–303. https://doi.org/10/df2hhd.

Riera-Guasp, Martin, Jose A. Antonino-Daviu, and Gérard-André Capolino. 2015. "Advances in Electrical Machine, Power Electronic, and Drive Condition Monitoring and Fault Detection: State of the Art." *IEEE Transactions on Industrial Electronics* 62 (3): 1746–59. https://doi.org/10/gf3tk7.

Salakhutdinov, Ruslan, and Geoffrey Hinton. 2009. "Deep Boltzmann Machines." Proceedings of the Twelfth International Conference on Artificial Intelligence and Statistics, 448–55. PMLR. https://proceedings.mlr.press/v5/salakhutdinov09a.html.

Sun, Wenjun, Siyu Shao, Rui Zhao, Ruqiang Yan, Xingwu Zhang, and Xuefeng Chen. 2016. "A Sparse Auto-Encoder-Based Deep Neural Network Approach for Induction Motor Faults Classification." *Measurement* 89 (July): 171–78. https://doi.org/10/gfx3vt.

Sun, Jianwen, Alexander Steinecker, and Philipp Glocker. 2014. "Application of Deep Belief Networks for Precision Mechanism Quality Inspection." In *Precision Assembly Technologies and Systems*, edited by Svetan Ratchev, 87–93. IFIP Advances in Information and Communication Technology. Berlin, Heidelberg: Springer. https://doi.org/10/gstnpp.

Sun, Jianwen, Reto Wyss, Alexander Steinecker, and Philipp Glocker. 2014. "Automated Fault Detection Using Deep Belief Networks for the Quality Inspection of Electromotors." *Tm—Technisches Messen* 81 (5): 255–63. https://doi.org/10/ggpkdq.

Szegedy, Christian, Wei Liu, Yangqing Jia, Pierre Sermanet, Scott Reed, Dragomir Anguelov, Dumitru Erhan, Vincent Vanhoucke, and Andrew Rabinovich. 2015. "Going Deeper With Convolutions." Proceedings of the IEEE Conference on Computer Vision and Pattern Recognition, 1–9. https://www.cv-foundation.org/openaccess/content_cvpr_2015/html/Szegedy_Going_Deeper_With_2015_CVPR_paper.html.

Tai, Kai Sheng, Richard Socher, and Christopher D. Manning. 2015. "Improved Semantic Representations from Tree-Structured Long Short-Term Memory Networks." arXiv. https://doi.org/10.48550/arXiv.1503.00075.

Tamilselvan, Prasanna, and Pingfeng Wang. 2013. "Failure Diagnosis Using Deep Belief Learning Based Health State Classification." *Reliability Engineering & System Safety* 115 (July): 124–35. https://doi.org/10/f4wcpc.

Tran, Van Tung, Faisal AlThobiani, and Andrew Ball. 2014. "An Approach to Fault Diagnosis of Reciprocating Compressor Valves Using Teager–Kaiser Energy Operator and Deep Belief Networks." *Expert Systems With Applications* 41 (9): 4113–22. https://doi.org/10/gfx3tr.

Wang, Jinjiang, Robert X. Gao, and Ruqiang Yan. 2014. "Multi-Scale Enveloping Order Spectrogram for Rotating Machine Health Diagnosis." *Mechanical Systems and Signal Processing* 46 (1): 28–44. https://doi.org/10/f5wpjv.

Wang, Yingmin, Fujun Zhang, Tao Cui, and Jinlong Zhou. 2016. "Fault Diagnosis for Manifold Absolute Pressure Sensor(MAP) of Diesel Engine Based on Elman Neural Network Observer." *Chinese Journal of Mechanical Engineering* 29 (2): 386–95. https://doi.org/10/gstnnf.

Xiong, Caiming, Stephen Merity, and Richard Socher. 2016. "Dynamic Memory Networks for Visual and Textual Question Answering." Proceedings of 33rd International Conference on Machine Learning, 2397–2406. PMLR. https://proceedings.mlr.press/v48/xiong16.html.

Yan, Ruqiang, Robert X. Gao, and Xuefeng Chen. 2014. "Wavelets for Fault Diagnosis of Rotary Machines: A Review with Applications." *Signal Processing* 96 (March): 1–15. https://doi.org/10/f5qcfn.

Yang, Xueliang, Ruqiang Yan, and Robert X Gao. 2015. "Induction Motor Fault Diagnosis Using Multiple Class Feature Selection." 2015 IEEE International Instrumentation and Measurement Technology Conference (I2MTC) Proceedings, 256–60. https://doi.org/10/gstnp5.

Zhang, Meijun, Jian Tang, Xiaoming Zhang, and Jiaojiao Zhang. 2016. "Intelligent Diagnosis of Short Hydraulic Signal Based on Improved EEMD and SVM With Few Low-Dimensional Training Samples." *Chinese Journal of Mechanical Engineering* 29 (2): 396–405. https://doi.org/10/gsqcfs.

CHAPTER 4

Convolutional Neural Networks for Intelligent Fault Diagnosis

INTRODUCTION

Deep neural networks (DNNs) like autoencoder-based DNNs (Sun et al. 2016) and deep belief networks (Shao et al. 2017) have been proven to be effective for intelligent fault diagnosis. However, these fully connected network structures are unable to learn the local and invariant features, which is critical for effective feature extraction from varied vibration signals. Convolutional neural networks (CNNs) (Le Cun et al. 1990) adopting sparse local connections between layers are naturally equipped with the intrinsic inductive bias of locality and translation equivariance. Consequently, CNNs are appropriate for learning from massive data and extracting local and invariant features. CNN is a biologically inspired variant of neural network, which has been proven to be successful in computer vision tasks (Krizhevsky, Sutskever, and Hinton 2012; Zeiler and Fergus 2014). Compared to vision tasks that address two-dimensional (2D) data, CNN can also be applied to time-series data as in LeCun and Bengio (1995) and Ossama et al. (2014), where the operation of a filter over sub-regions is performed along the time axis or the frequency axis to derive a feature map. Therefore, benefiting from CNNs' intrinsic inductive bias of locality and translation equivariance, CNNs have been widely investigated in fault diagnosis both on 2D image (Wang Peng, Yan, and Gao 2017; Lu, Wang, and Zhou 2017) and 1D time-series signal (Ince et al. 2016; Jing et al. 2017) and have shown satisfactory performance.

Next, the basic structure of CNN and a CNN-based intelligent fault diagnosis method (CDFL) (Sun et al. 2017) are introduced, respectively.

The rest of this chapter is organized as follows: basic modules of CNN, presentation of a CNN-based intelligent fault diagnosis method, datasets, evaluation results, and further discussions.

DOI: 10.1201/9781003474463-5

BASIC MODULES OF CNN

This section is going to present the basic structure of CNNs. As shown in Figure 4.1, the basic structure of CNN is mainly composed of convolutional layer, pooling layer, and fully connected layer. Considering that 2D-CNN has been illustrated extensively in previous research compared to 1D-CNN, in this section, the mathematical details behind 1D-CNN are given.

Convolutional Layer

The input of the convolutional layer is a sequential signal with length D. The convolutional filter with a window size of N is applied to the whole input signal. Specifically, the convolution operation in the convolutional layer can be defined as a multiply operation between a filter vector w, $w \in R^N$, and a concatenation vector representation $x_{i:i+N-1}$, which is given as:

$$x_{i:i+N-1} = x_i \oplus x_{i+1} \oplus \cdots \oplus x_{i+N-1}, \tag{4.1}$$

where $x_{i:i+N-1}$ can represent a window of N length sequential signal starting from the point indexed as i to the point indexed as $i + N - 1$, and $\oplus$ concatenates each data point into a longer embedding. In addition, a bias term b is also added into the convolution operation so that the final operation is given as the following equation:

$$z_i = \phi\left(w^T x_{i:i+N-1} + b\right), \tag{4.2}$$

where w^T denotes the matrix transpose of a matrix w and ϕ is a nonlinear activation function such as Sigmoid, Hyperbolic Tangent, or ReLu. Each specific vector w can be regarded as a convolutional filter. The single value z_i can be regarded as the learned feature for the window of sequences in the corresponding filter. In addition, the width of the filter, i.e., N can be varied.

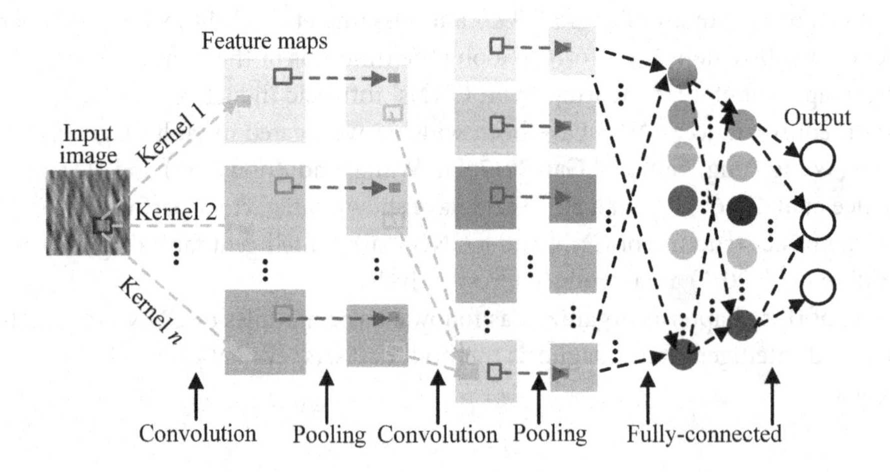

FIGURE 4.1 Basic structure of convolutional neural networks.

Since each raw sequence with a length D has multi-windows of sequences $\{x_{1:N}, x_{2:N+1}, \ldots, x_{D-N+1:D}\}$, a feature map can be obtained and the length of the feature map is determined based on the length of raw sequence D and the filter window size N. The feature map of the j-th filter can be denoted as:

$$z_j = \left[z_j^1, z_j^2, \ldots, z_j^{D-N+1}\right], \tag{4.3}$$

where the index j denotes the j-th filter.

Pooling Layer

A pooling layer is applied to the feature maps generated by the convolutional layer. The purpose of the pooling layer is twofold. First, the pooling operation will extract the most important and invariant information considering the single feature is irrelevant to the positions of activations in each feature map. Second, the pooling operation can reduce the feature dimensionality and guarantee the capability of addressing input signals with variant lengths. Normally, average-pooling and max-pooling are widely adopted. Here, the average-pooling function is applied as an example.

An average-pooling with a pooling length g is conducted in the feature maps obtained by the convolutional layer. Then, the single feature corresponding to the filter can be obtained as:

$$h_j = \left[h_j^1, h_j^2, \ldots, h_j^s\right], \tag{4.4}$$

$$h_j^t = \frac{1}{g}\sum_{i=1}^{g} z_j^{(t-1)g+i}, \tag{4.5}$$

where h_j is an s-dimensional vector, which is the output of pooling layer applied to the j-th feature map and $s = (D - N + 1)/g$ with D, N, and g being the length of input, filter size, and pooling size, respectively. Generally, multiple filters are applied with different weights and different window sizes to derive a feature vector. Assuming there are K filters, the output of the pooling layer is the concatenation of all outputs of filters after pooling, as shown in the following equation:

$$\mathbf{q} = [h_1, h_2, \ldots, h_K]. \tag{4.6}$$

Fully Connected Layer

All the neuron nodes of the fully connected layer are connected to all the neuron nodes in the feature map of the upper pooling layer output. The output of fully connected layer can be written as:

$$g(q) = \phi(w_1 q + b_1). \tag{4.7}$$

where q is the input of the fully connected layer, $g(q)$ is the output of the fully connected layer, w_1 is the connected weights, b_1 is the additive bias, and ϕ is the activation function.

Classification is usually carried out on the final output layer of the fully connected layers by using the Softmax classifier (Liao et al. 2015), which is an extension of the Logistic classifier, and mainly solves multi-class problems. Assuming an input sample x with the corresponding label y, the probability of the sample judged as a class j is $p(y = j|x)$. So, for a c classifier, the output will be a c-dimensional vector as given in Eq. (4.8).

$$g_\theta\left(x^{(i)}\right)=\begin{bmatrix} p\left(y^{(i)}=1\middle|x^{(i)};\theta\right) \\ p\left(y^{(i)}=2\middle|x^{(i)};\theta\right) \\ \vdots \\ p\left(y^{(i)}=c\middle|x^{(i)};\theta\right) \end{bmatrix}=\frac{1}{\sum_{j=1}^{c} e^{\theta_j^T x^{(i)}}}\begin{bmatrix} e^{\theta_1^T x^{(i)}} \\ e^{\theta_2^T x^{(i)}} \\ \vdots \\ e^{\theta_c^T x^{(i)}} \end{bmatrix}. \tag{4.8}$$

Where θ is the model parameter, $1/\Sigma_{j=1}^{c} e^{\theta_j^T x^{(i)}}$ is a normalization function that normalizes the probability distribution so that all probabilities sum total to 1.

The cost function $J(\theta)$ of Softmax classifier is minimized by multiple iterations using the gradient descent method, and thereby completing the network training. $J(\theta)$ is expressed as:

$$J(\theta)=-\frac{1}{m}\left[\sum_{i=1}^{m}\sum_{j=1}^{c}1\left\{y^{(i)}=j\right\}\log\frac{e^{\theta_j^T x^{(i)}}}{\sum_{l=1}^{c} e^{\theta_l^T x^{(i)}}}\right]. \tag{4.9}$$

where $1\{\bullet\}$ is an indicator function, namely, when the value in brace is true, the function value is 1, otherwise, the value is 0. And m indicates the number of data samples.

Thus, after an input sample is alternately propagated through multiple convolutional and pooling layers, the extracted features will be forwarded to the fully connected layers and finally classified by Softmax classifier.

CNN-BASED INTELLIGENT FAULT DIAGNOSIS

In this section, a 1D CNN-based intelligent fault diagnosis framework is presented, where a feature learning scheme named as convolutional discriminative feature learning (CDFL) is proposed. CDFL adopts a specific neural network for filter weights learning instead of unsupervised methods in the original unsupervised CNN for the robust and discriminative representations.

Discriminative Learning

A discriminative weight learning scheme is proposed by applying a neural network with a specific structure and transferring the parameters of hidden layers into the convolutional layer in unsupervised CNN. It has been indicated by Zhao et al. (2010) and Yang and Li (2015) that Back Propagation Neural Network (BPNN) is of good ability to discriminate motor conditions. Therefore, the proposed CDFL utilizes BPNN as a discriminative learning method to learn the parameters of the convolutional layer. Due to the discriminative pre-training, a fast and effective convolutional pooling architecture can be performed to learn discriminative and invariant features automatically from raw vibration data of the induction motor, which does not require large-scale training samples.

To transfer the parameters of BPNN into the parameters of the convolutional pooling architecture, the structure of BPNN is designed as follows: BPNN consists of input layer, hidden layer, and output layer. Sigmoid functions are used as the activation functions for both the hidden and output layers. The sizes of BPNN's input layer and hidden layer are set to be the local filters' window size N and the number of filters K, respectively. After training, the parameters between the input layer and hidden layer of the BPNN can be directly used as convolutional layer's parameter. The learned transformation matrix in BPNN can be denoted as W with a size of $N \times K$. Then, each column in the matrix W can be regarded as one filter's weight.

To train the specifically structured BPNN, the templates of vibration signals x^i and their labels y_i should be constructed as follows: $X_0 = (x^i, y_i), x^i \in R^N, i = 1,2,\ldots,M$, where M is the number of the template samples to train BPNN and the length of each data sample is N corresponding to the input layer size of the BPNN. Since each x^i is subsampled from the original sequential data, its label y_i is set to be the same as the one of the original sequential data. Since each x^i is subsampled, M usually exceeds the size of original training data samples. Therefore, this framework can be regarded as a kind of data augmentation, which can relieve overfitting problem (Gan et al. 2015).

SVM Classification

Compared to supervised CNN where a Softmax classifier is applied as the top layer, the model adopted a more robust classifier: support vector machine (SVM). The learned features are fed into the kernel SVM classifier through nonlinear transform (Yin et al. 2015). Here, the Gaussian radial basis function (RBF) is chosen as the SVM's kernel:

$$K(O_i, O_j) = \exp\left(-\frac{\|O_i - O_j\|^2}{2\sigma^2}\right) \tag{4.10}$$

where O_i and O_j denote the learned representations of the i-th and j-th data samples, and σ is the Gaussian kernel parameter.

In the case of multi-class predictions, a one-against-one strategy where multiple binary classifiers are trained using the data from two classes is adopted. The trained classifier is then applied to the testing dataset to predict their fault categories.

The LIBSVM tools are used for the implementation of the SVM classification (Chang and Lin 2011), where the regularization term C that controls norms of model parameters and the Gaussian kernel parameter σ of the RBF-based SVM classifier are determined by cross-validation.

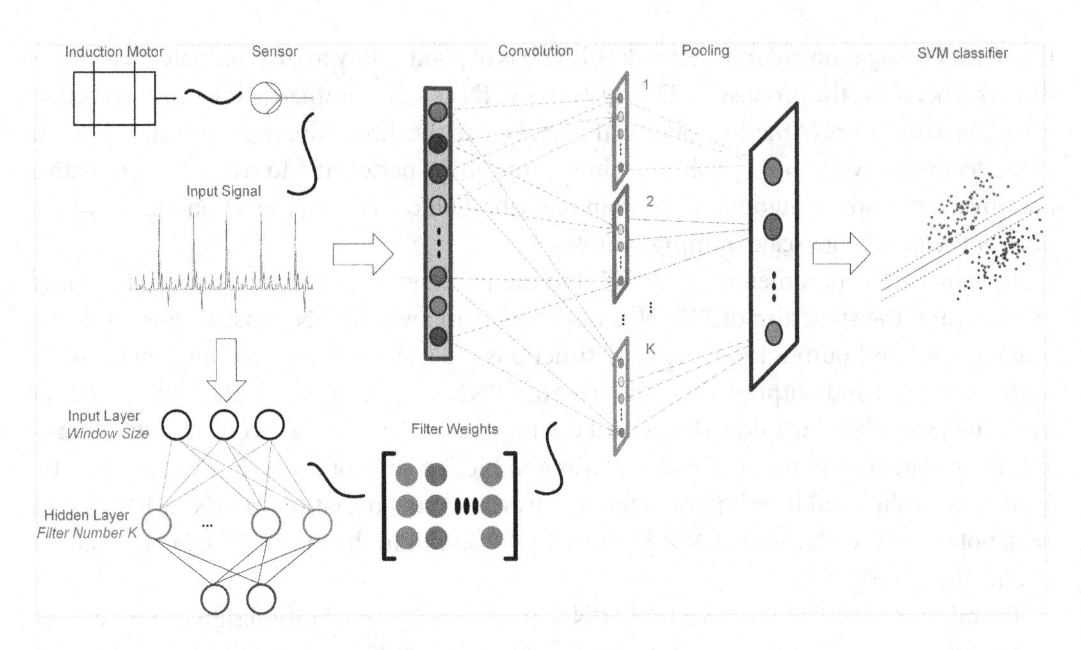

FIGURE 4.2 The framework of CDFL method.

CNN-Based Intelligent Fault Diagnosis

In the proposed CDFL framework, a convolutional pooling architecture is first designed with K local filters of N length window size. Then, a BPNN is constructed, which is a three-layer structure where the sizes of input layer, hidden layer, and output layer are filter window size, filter number, and number of fault conditions, respectively. Therefore, the learning templates are prepared for discriminative learning, whose lengths are set to be the same as filters' window size. These templates with their fault labels are used to train the BPNN. The parameters between the input layer and hidden layer are transferred into the one for convolutional layer. Then, the convolutional pooling architecture with the transferred parameters is applied to the raw data to learn new representations. Finally, the new representations are fed into SVM for classification to predict fault categories. As a summary, BPNN is used to learn the model parameters and then the learned parameters are adopted for the local filters realizing the convolutional discriminative feature learning. Then the learned representation can be fed for the classification of SVM. The whole procedure for induction motor fault diagnosis via the proposed CDFL method is illustrated in Figure 4.2.

EXPERIMENTAL STUDIES

In this section, we test the induction motor dataset to verify the performance of the CDFL method for comparative studies.

Induction Motor Dataset

The machine fault simulator is illustrated in Figure 4.3. The machine fault simulator is driven by a 1/2-hp induction motor (Model: Marathon Motors 56T34F5301J). A triaxial ring-style industrial accelerometer (Model: PCB 604B31) with ±50 g measurement range at 100 mv/g sensitivity is mounted on the surface of the induction motor near to the shaft end

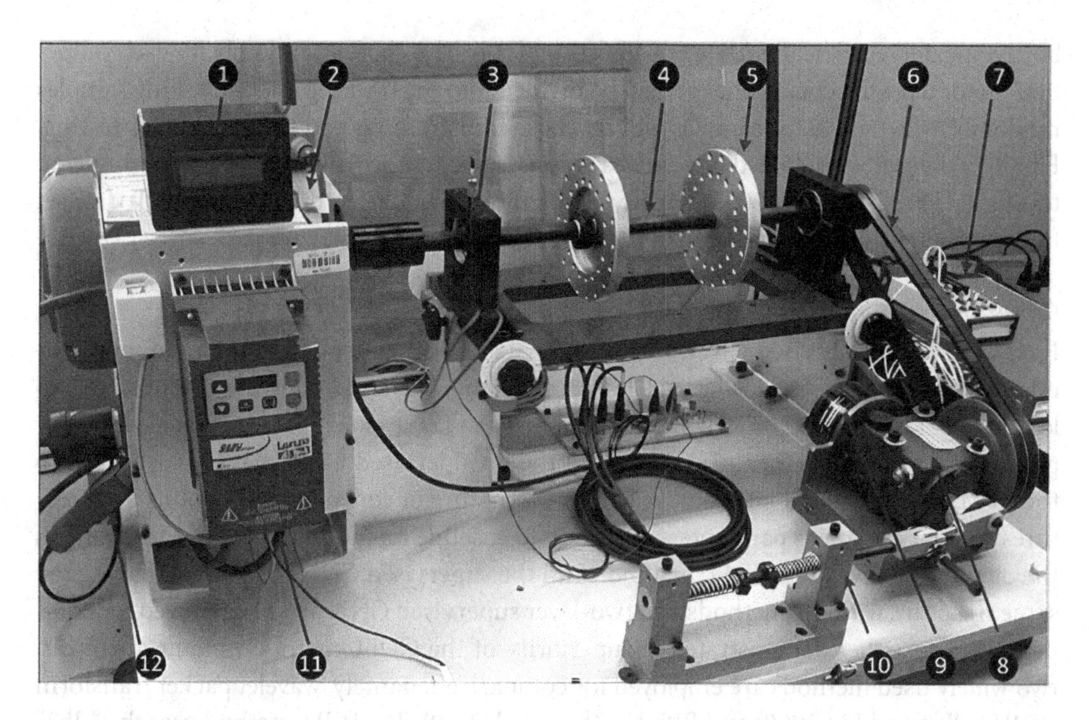

FIGURE 4.3 Experimental setup: (1) tachometer, (2) induction motor, (3) bearing, (4) shaft, (5) load disc, (6) belt, (7) data acquisition board, (8) bevel gearbox, (9) magnetic load, (10) reciprocating mechanism, (11) variable speed controller, and (12) current probe.

to measure vibration signals under six different conditions. The *Y*-axial vibration signal of the simulator under the power supply frequency of 60 Hz is collected with the sampling frequency of 20 kHz for the trial. The descriptions of different induction motor conditions are listed in Table 4.1.

Experimental Description

Data Preprocessing and Splitting

After the vibration signals of six different conditions have been acquired, the data samples with their corresponding labels (working conditions) can be constructed from these signals. To construct training and testing data, firstly a window containing 799 sampling points is regarded as one data sample for training and testing, which means each data sample for training or testing is represented by 799 time-steps, which almost consists two complete periodic signals. In the following section, the influence of number of sampling points will

TABLE 4.1 The Description of the Class Labels of the Induction Motor

Condition		Description
Health	Normal motor	Healthy, no defect
SSTM	Stator winding defect	3 turns shorted in stator winding
UBM	Unbalanced rotor	Unbalance caused by 3 added washers on the rotor
RMAM	Defective bearing	Inner race defect bearing in the shaft end
BRB	Broken bar	Three broken rotor bars
BRM	Bowed rotor	Rotor bent in center 0.01″

be investigated empirically. Then 600 data samples will be chosen randomly for each induction motor working condition, and will be divided into the training and the testing samples randomly, in which 2/3rd samples will be chosen for training and the rest 1/3 for testing. Finally, a dataset with 3600 data samples is constructed, in which numbers of training and testing data are 2400 and 1200, respectively. All the individual features for the whole data set in experiments are all normalized to the range between −1 and 1.

Comparison Approaches

First, several unsupervised CNNs including sparse autoencoder (SAE) and random methods are considered here. SAE-based CNN (Fan Zhang, Du, and Zhang 2014) adopts SAE to learn the parameters of CNN, while random method (Saxe et al. 2011) generated the CNN parameters randomly. Then, the outputs of the unsupervised CNN methods are fed into the SVM classifier. For a fair comparison, the structure of convolutional pooling architectures and the classifier's parameters in these two baseline methods are set to be the same as the ones in the proposed CDFL method. And the supervised CNN (one layer) sharing the same structure with the methods and two-layer supervised CNN are all employed for comparison. To further demonstrate the superiority of the CDFL method in fault diagnosis, two widely used methods are employed for comparison, namely, wavelet packet transform (WPT) (Yen and Lin 2000) and BPNN (Zhao et al. 2010). The WPT method uses the "db2" wavelet to decompose the training samples into seven layers and extract the wavelet packet node energy on all levels. The features finally are obtained as concatenation vectors of different energy values from all seven layers. Then the learned vectors with a dimensionality of 254 are used for training of the SVM classifier. The BPNN contains one hidden layer with 100 hidden nodes. In addition, to highlight the effectiveness of the proposed method, deep learning method like the DNNs with single-layer SAE (Sun et al. 2016), is employed for comparison. DNN has the same number of hidden nodes as the BPNN, which is set to 100. For a fair comparison, the CNN with two layers has the same number and size of local filters as the first layer of the proposed model except for the pooling size, while the second convolution and pooling layer has 6 local filters with size of 51.

Training Details

The training protocols of BPNN, DNN, and CNNs all adopt the same iteration numbers as 4800 times. The training time is calculated from the first iteration to the end of the training process for the BPNN, DNN, and CNNs, and is measured including the feature learning and classification process of training data for the WPT and the CDFL method.

The simulation environment for all algorithms performed in the experiments is as follows: MATLAB R2011b software environment, Computer operating system Windows 7, the CPU frequency of 3.3 GHz, 8192MB RAM.

Results and Discussion

Results of Datasets

In the first set of the experiments, the effects of the local filters' window size and the pooling size in the convolutional pooling architecture are evaluated. It is noted that the reported results

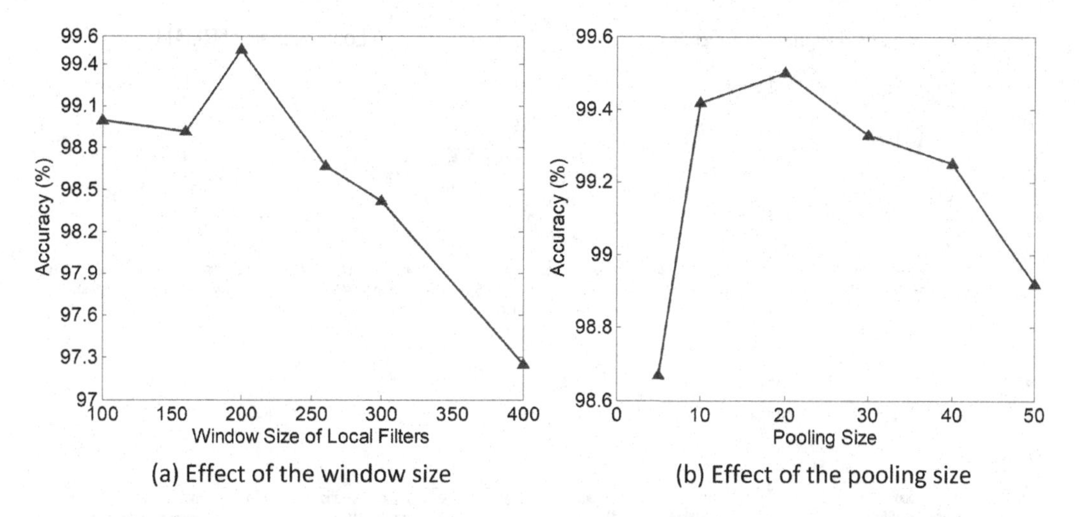

FIGURE 4.4 The effect on the performance of the window size (a) and the pooling size (b).

are all based on validation datasets, which are portion of training datasets. The local filters' window sizes are varied from 100 to 400, and templates' dimension and BPNN's structure change accordingly. Figure 4.4(a) shows the classification accuracies with different window sizes. It can be seen that the best classification performance was obtained at 200. The model can achieve the best performance under a moderate window size, since the filters with a small window size cannot capture the long-distance information in the sequential signal and the filters with a large window size introduce too many parameters. In the same way, different pooling size values were investigated. Figure 4.4(b) shows the classification performances with different pooling sizes. It is shown that the best classification performance can be obtained at a value close to 20 and the performance starts to deteriorate when the pooling size is more than 20. This indicates that proper pooling size can improve the performance of the proposed method, since a large pooling size may smooth some important features in the pooling operation so that some specific signals are lost. It should be noted that the use of local filters without average-pooling only gives a bad performance which corresponds to the scenario that pooling size is zero. Therefore, in the proposed CDFL method, the window size and pooling size are set to be 200 and 20, respectively.

To demonstrate that CDFL can derive invariant and robust features, different local regions of motor vibration signals are taken as input in CDFL to learn corresponding features. The raw local signal maps and the learned feature maps of fault signal broken rotor bar (BRB) and RMAM are shown in Figure 4.5. It can be seen that features of the same fault condition almost remain invariant although the input signal is different in time domain. It can be found from the feature maps that the learned features are robust and show some internal information about the vibration signal.

To reflect the advantages of the proposed discriminative learning method, comparison between the proposed method and two baseline unsupervised CNN methods including SAE and random ones is firstly presented. Figure 4.6 presents the classification accuracies of the three methods, which are calculated over the whole categories. In addition, variance

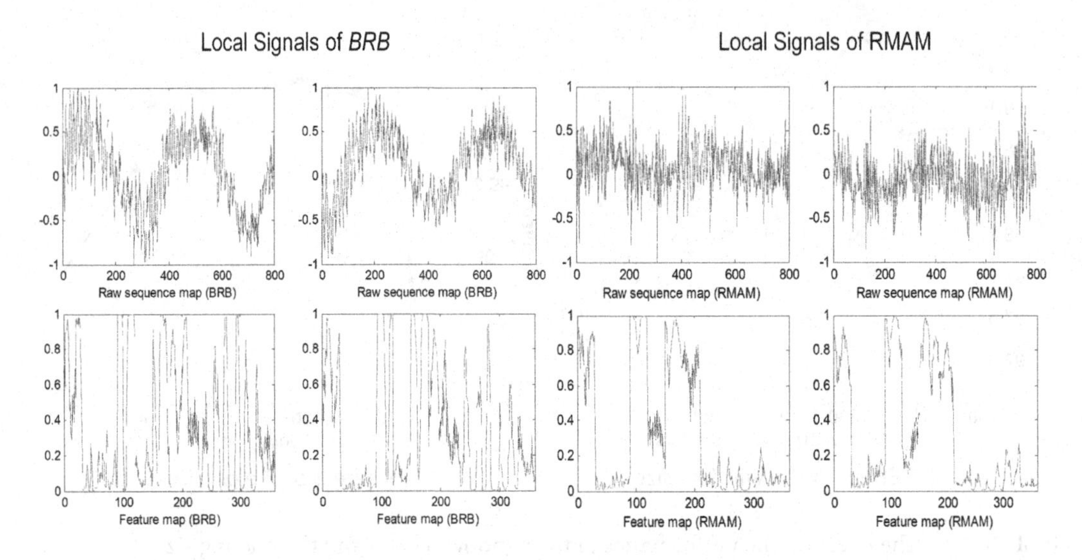

FIGURE 4.5 Local signal maps and the CDFL learned feature maps.

analysis is also conducted in which 30 randomized trials are analyzed and its average and variance values are reported. It can be seen that the proposed CDFL achieves the best performance including high accuracy and low variance. Therefore, the proposed discriminative learning is more effective for learning the filters' weights of the fast convolutional pooling architecture compared to these two unsupervised methods. It can be explained by the fact that the mapping function learned by the BPNN can give a discriminative guidance to the convolutional pooling architecture when generating features.

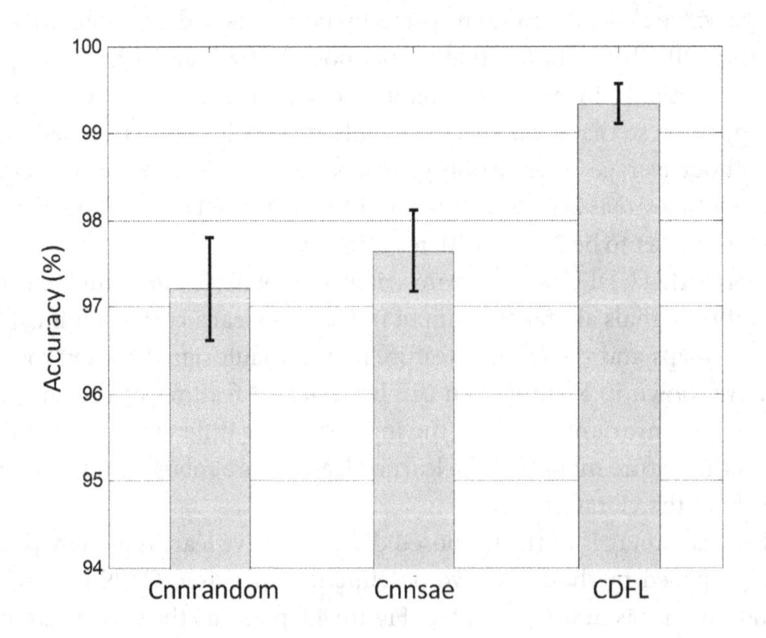

FIGURE 4.6 Comparison of three methods.

TABLE 4.2 Comprehensive Comparison of Classification Accuracy

Vy	Health	BRB	BRM	RMAM	SSTM	UBM	Average	Time
BPNN	74.20%	92.00%	82.80%	94.60%	82.30%	79.40%	84.22%	15 s
DNN-SAE	85.05%	97.90%	99.65%	99.85%	87.35%	96.35%	94.36%	40 s
CNN 1 layer	85.00%	83.50%	55.50%	99.00%	93.00%	98.00%	85.67%	250 s
CNN 2 layer	96.80%	99.90%	100.00%	100.00%	98.20%	99.90%	99.13%	500 s
WPT + SVM	96.50%	100%	100.00%	100.00%	94.50%	100.00%	98.50 %	43 s
CDFL + SVM	**97.80%**	**100.00%**	**100.00%**	**100.00%**	**98.45%**	**99.90%**	**99.36%**	**5 s**

The classification accuracy and training time of all compared methods are shown in Table 4.2. Compared to the one hidden layer neural networks including BPNN, DNN-SAE, and one-layer CNN, the proposed CDFL method achieves a higher classification accuracy. Especially, the performance gain achieved by the CDFL method compared to the supervised CNN with the same structure is significant as 14% in the average accuracy. It shows that in fault diagnosis, the proposed method that incorporates a novel pre-training of CNN parameters is more effective than traditional CNN. In addition, the proposed CDFL performs slightly better than the two-layer CNN, although the two-layer CNN is more complex than the CDFL method. When compared to the traditional CNNs, besides the higher classification accuracy, the training time of CDFL + SVM is much less. At the same time, compared to the state-of-the-art method, WPT, the CDFL method has shown better performance.

Analysis of Different Input Data Length for Our Models

In the above section, the input data length of all models is set to 799. And it is meaningful to investigate the influence of the different input data length, i.e., the number of sampling points, on classification performances of the proposed model. The proposed method is also compared with these two best competitive methods including WPT + SVM and two-layer CNN. Here, the input data length is varied from 399 to 1999 with a step size of 200. As shown in Figure 4.7, the performances of CDFL and WPT models increase with the input data length increasing in the given scale, while the CNN with two-layer structure decreases sharply when the input data length is larger than 1199. Considering the scale of the figure, the results for accuracy of two-layer CNN are not plotted when the input data length is larger than 1199. The results of the first two models may be explained by the fact that the input data with a large sampling point contains more information, and therefore the classification performance can be further improved with a large input data length. As for CNN models, there exists a serious overfitting problem when input data length is large. It may be explained by the fact that the limited training data sets become relatively small for the CNN models when the input data has a high dimensionality. CNN models need more training data and may need to adjust the filters' size. However, this does not hinder the performance of the proposed CDFL model, which further verify the effectiveness of the discriminative local filters learning scheme using BP-based neural network. At last, it is shown that the performance of CDFL model is slight worse than the one of WPT + SVM when the input data length is over 1000. It can be explained that the bigger the input data length is, the

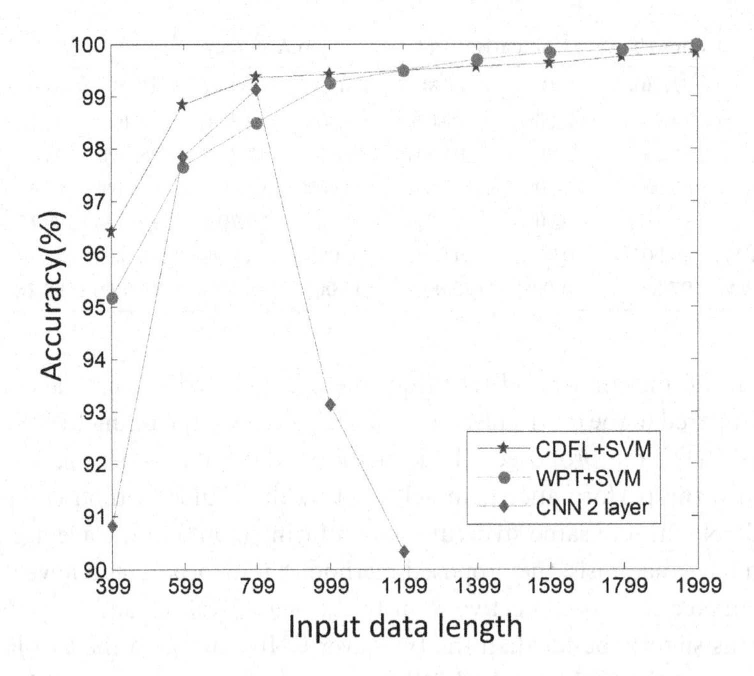

FIGURE 4.7 Accuracies of our models and two competitive models under different input data length.

dimension of features generated by CDFL is bigger while that generated by WPT remains the same. However, the proposed CDFL is able to outperform the WPT + SVM significantly under a small input data length. Therefore, it is more robust than the WPT + SVM.

Visualization of Learned Representation

In this section, the effectiveness of the proposed approach has been demonstrated qualitatively based on visualization of learned representation. A technique named "t-SNE" is adopted, which is an effective method to visualize high-dimensional data by mapping the data samples from their original feature space into a two- or three-dimensional space map (Van der Maaten and Hinton 2008). First, principal component analysis (PCA) is used to reduce the dimensionality of the feature data to 50. This can speed up the computation of pairwise distances between the data points and suppresses some noise without severely distorting the interpoint distances. Then the dimensionality reduction technique "t-SNE" is used to convert the 50-dimensional representation to a two-dimensional map. Here, different representations learned by different methods including raw data, DNN and the proposed CDFL are analyzed and different colors represent different working conditions. The resulting maps of raw data feature, DNN feature and CDFL feature as a scatterplot with two dimensions are shown in Figure 4.8, respectively. It can be seen that features of the same condition learned by CDFL are clustered the best while features of different conditions are separated well. In comparison, features learned by the DNN-SAE do not cluster well. For example, as shown in Figure 4.8 (a) and (b), the points of SSTM overlapped the points of Health so that an effective classification cannot be achieved. However, as shown in Figure 4.8 (c), the points of SSTM and Health are separated very well in the feature space learned by CDFL. It should be noted that since the final classification is conducted

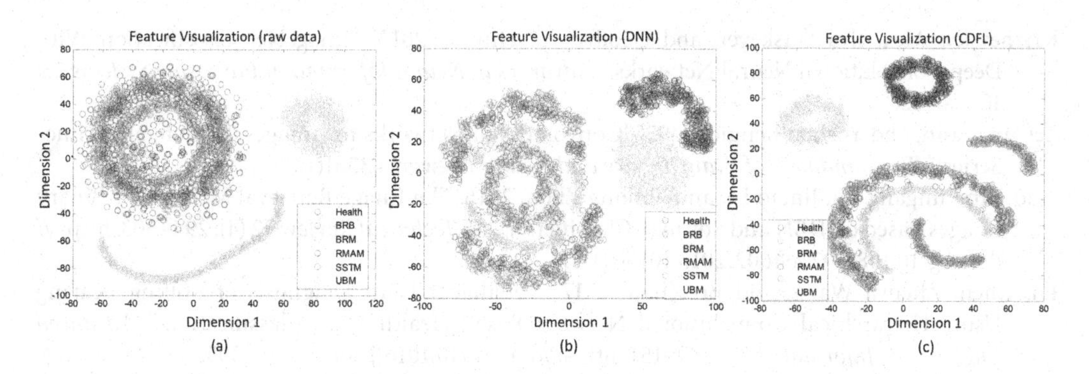

FIGURE 4.8 Feature visualization based on t-SNE.

in a high dimensionality space nonlinearly, several overlap among different classes in visualization is still consistent with the high accuracies in Table 4.2. The above visualization is able to qualitatively illustrate the effectiveness of the proposed CDFL for fault diagnosis.

CONCLUSION

The CDFL approach has been presented for induction motor fault diagnosis in this chapter. This approach can learn features directly from raw data to characterize different working conditions of the induction motors. When the local filters' weights are learned by BPNN, the fast convolutional pooling architecture can be built to extract the discriminative and invariant features from the raw vibration data. Then the SVM can classify the learned features successfully for induction motor fault diagnosis. Experimental studies have verified the CDFL's capability of extracting the discriminative and effective features from the fault signals. As compared to other state-of-the-art methods, the approach proposed in this chapter is more effective and robust for induction motor fault diagnosis. Further research is being conducted to verify the applicability of the presented approach for fault diagnosis of other types of machines.

REFERENCES

Chang, Chih-Chung, and Chih-Jen Lin. 2011. "LIBSVM: A Library for Support Vector Machines." *ACM Transactions on Intelligent Systems and Technology (TIST)* 2 (3): 1–27. https://doi.org/10.1145/1961189.1961199.

Gan, Zhe, Ricardo Henao, David Carlson, and Lawrence Carin. 2015. "Learning Deep Sigmoid Belief Networks With Data Augmentation." *Artificial Intelligence and Statistics, PMLR* 38: 268–276. http://proceedings.mlr.press/v38/gan15.

Ince, Turker, Serkan Kiranyaz, Levent Eren, Murat Askar, and Moncef Gabbouj. 2016. "Real-Time Motor Fault Detection by 1-D Convolutional Neural Networks." *IEEE Transactions on Industrial Electronics* 63 (11): 7067–7075. https://doi.org/10.1109/TIE.2016.2582729.

Le Cun, Y., B. Boser, J. S. Denker, D. Henderson, R. E. Howard, W. Hubbard, and L. D. Jackel. 1990. "Handwritten Digit Recognition With a Back-Propagation Network." *Advances in Neural Information Processing Systems* 2:396–404.

Jing, Luyang, Ming Zhao, Pin Li, and Xiaoqiang Xu. 2017. "A Convolutional Neural Network Based Feature Learning and Fault Diagnosis Method for the Condition Monitoring of Gearbox." *Measurement* 111: 1–10. https://doi.org/10.1016/j.measurement.2017.07.017.

Krizhevsky, Alex, Ilya Sutskever, and Geoffrey E. Hinton. 2012. "ImageNet Classification With Deep Convolutional Neural Networks." *Advances in Neural Information Processing Systems* 25: 1097–1105.

LeCun, Yann, and Yoshua Bengio. 1995. "Convolutional Networks for Images, Speech, and Time-Series." *The Handbook of Brain Theory and Neural Networks* 3361(10).

Liao, Bin, Jungang Xu, Jintao Lv, and Shilong Zhou. 2015. "An Image Retrieval Method for Binary Images Based on DBN and Softmax Classifier." *IETE Technical Review* 32 (4): 294–303. https://doi.org/10.1080/02564602.2015.1015631.

Lu, Chen, Zhenya Wang, and Bo Zhou. 2017. "Intelligent Fault Diagnosis of Rolling Bearing Using Hierarchical Convolutional Network Based Health State Classification." *Advanced Engineering Informatics* 32: 139–151. https://doi.org/10.1016/j.aei.2017.02.005.

Ossama, Abdel-Hamid, Abdel-rahman Mohamed, Hui Jiang, Li Deng, Gerald Penn, and Dong Yu. 2014. "Convolutional Neural Networks for Speech Recognition." *IEEE/ACM Transactions on Audio, Speech, and Language Processing* 22 (10): 1533–1545. https://doi.org/10.1109/TASLP.2014.2339736.

Saxe, Andrew M., Pang Wei Koh, Zhenghao Chen, Maneesh Bhand, Bipin Suresh, and Andrew Y. Ng. 2011. "On random weights and unsupervised feature learning." In NIPS Workshop on Deep Learning and Unsupervised Feature Learning: 1089–1096.

Shao, Siyu, Wenjun Sun, Ruqiang Yan, Peng Wang, and Robert X Gao. 2017. "A Deep Learning Approach for Fault Diagnosis of Induction Motors in Manufacturing." *Chinese Journal of Mechanical Engineering* 30: 1347–1356. https://doi.org/10.1007/s10033-017-0189-y.

Sun, Wenjun, Siyu Shao, Rui Zhao, Ruqiang Yan, Xingwu Zhang, and Xuefeng Chen. 2016. "A Sparse Auto-Encoder-Based Deep Neural Network Approach for Induction Motor Faults Classification." *Measurement* 89: 171–178. https://doi.org/10.1016/j.measurement.2016.04.007.

Sun, Wenjun, Rui Zhao, Ruqiang Yan, Siyu Shao, and Xuefeng Chen. 2017. "Convolutional Discriminative Feature Learning for Induction Motor Fault Diagnosis." *IEEE Transactions on Industrial Informatics* 13 (3): 1350–1359. https://doi.org/10.1109/TII.2017.2672988.

Van der Maaten, Laurens, and Geoffrey Hinton. 2008. "Visualizing Data Using t-SNE." *Journal of Machine Learning Research* 9 (11): 2579–2605.

Wang Peng, Ananya, Ruqiang Yan, and Robert X. Gao. 2017. "Virtualization and Deep Recognition for System Fault Classification." *Journal of Manufacturing Systems* 44: 310–316. https://doi.org/10.1016/j.jmsy.2017.04.012.

Yang, Tan, and Xiangru Li. 2015. "An Autoencoder of Stellar Spectra and Its Application in Automatically Estimating Atmospheric Parameters." *Monthly Notices of the Royal Astronomical Society* 452 (1): 158–168. https://doi.org/10.1093/mnras/stv1210.

Yen, Gary G., and K-C. Lin. 2000. "Wavelet Packet Feature Extraction for Vibration Monitoring." *IEEE Transactions on Industrial Electronics* 47 (3): 650–667. https://doi.org/10.1109/41.847906.

Yin, Zuyu, Jianxing Liu, Minjia Krueger, and Huijun Gao. 2015. "Introduction of SVM Algorithms and Recent Applications About Fault Diagnosis and Other Aspects." 2015 IEEE 13th International Conference on Industrial Informatics (INDIN) IEEE, 550–555. https://doi.org/10.1109/INDIN.2015.7281793

Zeiler, Matthew D., and Rob Fergus. 2014. "Visualizing and Understanding Convolutional Networks." Computer Vision–ECCV 2014: 13th European Conference, Zurich, Switzerland, September 6–12, 2014, Proceedings, Part I 13 (pp. 818–833). Springer International Publishing.

Zhang, Fan, Bo Du, and Liangpei Zhang. 2014. "Saliency-Guided Unsupervised Feature Learning for Scene Classification." *IEEE Transactions on Geoscience and Remote Sensing* 53 (4): 2175–2184. https://doi.org/10.1109/TGRS.2014.2357078.

Zhao, Xiaodong, Xinliang Tang, Juan Zhao, and Yubin Zhang. 2010. "Fault Diagnosis of Asynchronous Induction Motor Based on BP Neural Network." 2010 International Conference on Measuring Technology and Mechatronics Automation 236–239. Changsha, China. https://doi.org/10.1109/ICMTMA.2010.417

II

Advanced Topics of Deep Learning-Enabled Intelligent Fault Diagnosis

CHAPTER 5

Data Augmentation for Intelligent Fault Diagnosis

INTRODUCTION

For deep learning-based fault diagnosis methods, training data is an important factor that affects the performance of prediction model. Generally, a deep neural network contains multiple hidden layers, and the number of free parameters that need to be trained is enormous. In order to achieve accurate predictions, the deep architecture need to be well-trained, and fully training a large network usually requires abundant balanced data. However, in practice, training samples across different machine states often exhibit an imbalance. For example, for a mechanical system under operation, it is working under normal circumstance most of the time and the collected sensor data that represent positive training samples are sufficient, while machine operating in fault is infrequent and corresponding collected samples are limited compared to positive samples. Therefore, there is an imbalance between positive training samples and fault samples. Suffered from limited training data, especially unbalanced data, training a deep model to achieve accurate prediction of machine conditions is relatively difficult.

Generative adversarial networks (GANs) offer an alternative when facing the issue of unbalanced data classification by generating data for minor classes. GANs have shown prominent capability in producing realistic looking data and have been successfully applied in image generation. It can be used for data augmentation by generating artificial data that are similar to original data and thereby enriching training dataset. Moreover, the generative architecture models data generation process and therefore helps to understand the original data distribution, providing a new perspective for creating predictive systems in machine fault diagnosis tasks.

GAN was first introduced as a framework for generation of artificial images (Goodfellow et al. 2014) and has the capability to produce convincing image samples. However, the original GAN model faced challenges with training instability, requiring

DOI: 10.1201/9781003474463-7

the implementation of elaborate regularization techniques to achieve desirable performance. Therefore, more researches have been carried out in regard to model stability and generation quality. Wasserstein GAN (WGAN) leverages the Wasserstein distance to form a new loss function which contributes to better model stability than the original, and it relates loss function to the quality of generated images (Arjovsky, Chintala, and Bottou 2017), and an improved training strategy for WGAN is proposed to insure stable training of various GAN architecture (Gulrajani et al. 2017). A structure which combined deep convolutional generative adversarial networks (DCGANs) with certain architectural constraints is designed to learn a hierarchy of representations from object parts to scenes for unsupervised learning (Radford, Metz, and Chintala 2016). Semi-supervised learning using GAN is introduced to produce class labels in discriminator network and improve generated samples quality (Odena 2016). Afterward, a new variant GAN, called auxiliary classifier GAN (ACGAN), using label information was proposed to generate high-resolution images and achieve desirable performance in classification tasks (Odena, Olah, and Shlens 2016).

GANs and their various iterations have demonstrated their effectiveness in generating images, and more recent applications have extended their potential to the generation of artificial audio (Donahue, McAuley, and Puckette 2018) and electroencephalographic (EEG) brain signals (Hartmann, Schirrmeister, and Ball 2018), showcasing their ability to generate time-series data. However, there has been limited exploration in generating raw data of sensor signals. Moreover, existing research on evaluating GANs primarily relies on visual estimation of sample quality, which may not be suitable for scenarios involving sensor signals. As a result, we represent a pioneering effort in utilizing the ACGAN architecture to generate mechanical sensor signals for data augmentation in conjunction with subsequent fault classification. In addition, we propose an evaluation system designed to assess the quality of the generated samples.

In this chapter, an ACGAN architecture based on one-dimensional convolutional layers (1D-CNN) is constructed to learn features from limited training data and generate realistic sensor data, and the high-quality generated samples can be used in further applications in machine fault diagnosis. The 1D-CNN is adopted as the building block of both generator and discriminator to leverage its capability of learning local and hierarchical representations from raw data which are beneficial for classification tasks and allow for feature interpretability. Category labels are added in both generator and discriminator to help accelerate model training and batch normalization technique is performed in generator to overcome the problem of gradient vanishing and therefore avoid overfitting. To evaluate the performance of the generative model, a comprehensive set of metrics is employed to assess the quality of the generated samples both quantitatively and visually. The experimental results presented in this chapter showcase the data generation capabilities of the proposed framework using an induction motor dataset. These results not only highlight the effectiveness of the framework in generating synthetic data but also provide classification outcomes using the artificially augmented dataset.

The rest of this chapter is organized as follows: theoretical background, detailed GAN-based data augmentation for intelligent fault diagnosis method, experimental studies, and conclusion.

THEORETICAL BACKGROUND

This section is going to provide theoretical background of GANs, ACGANs, and similarity evaluation criteria.

Generative Adversarial Networks

A regular GAN consists of two parts which are trained in opposition to each other, the generator G and the discriminator D, as shown in Figure 5.1(a). The main thought behind GAN is using adversarial networks to improve the quality of generated data. The generator is trained to produce realistic synthesized data $x_{generated} = G(z)$ from a random noise vector z, trying to fool the discriminator so that $x_{generated}$ would not be recognized as generated samples. The generated distribution is denoted as P_g. On the contrary, the discriminator takes both the real training data and fake samples generated from generator as the input, and then it is trained to distinguish between generated samples and real data. Discriminator outputs the probability that certain samples correspond to possible data sources. The loss function of discriminator L can be defined as:

$$\begin{aligned} L &= E_{x\sim P_{data}}[\log p(s = real \mid x_{real})] + E_{z\sim P(z)}[\log p(s = generated \mid x_{generated})] \\ &= E_{x\sim P_{data}}[\log(D(x))] + E_{z\sim P(z)}[\log(1 - D(G(z)))] \end{aligned} \tag{5.1}$$

where P_{data} is the real data distribution, $P(z)$ is a prior distribution on noise vector z, $D(x)$ denotes the probability that x comes from the real data rather than generated data, $E_{x\sim P_{data}}$ represents the expectation of x from real data distribution P_{data}, and $E_{z\sim P(z)}$ is the expectation of z sampled from noise. For discriminator, the goal of training is to maximize the loss function which means maximizing the log-likelihood with correct sample source; while for generator, the training goal is to minimize the second term in Eq. (5.1) to confuse the discriminator. Therefore, the goal of the whole GAN architecture can be summarized as:

$$Goal = \arg \min_G \max_D L(G, D) \tag{5.2}$$

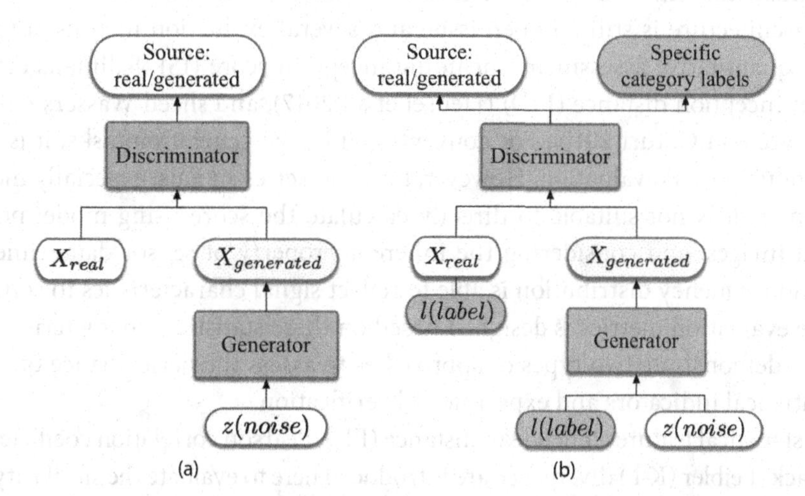

FIGURE 5.1 Typical architecture of (a) regular GAN, and (b) auxiliary classifier GAN.

Updating model parameters based on the objective function is able to train GAN by stochastic gradient descent (SGD) and achieve proper pair of discriminator and generator.

Auxiliary Classifier Generative Adversarial Networks

A variant architecture of the regular GAN is achieved by leveraging additional class labels for both discriminator and generator, as shown in Figure 5.1(b). Class conditional generation process is able to improve the quality of generated data. In addition, the discriminator is combined with an auxiliary part to output specific class labels so that the improved discriminator can not only recognize data source, but also differentiate among various classes. This variant that combines class conditional architecture and auxiliary network for classification is called auxiliary classifier generative adversarial network (ACGAN).

Compared with regular GANs, ACGANs are able to generate high-quality data and provide label information simultaneously. Mathematically, generator leverages both noise z and label l to produce artificial data samples $x_{generated} = G(z, l)$, and discriminator outputs probability on data sources and class labels. Therefore, the objective function contains two separate log-likelihoods corresponding to correct data source and correct class label, shown as:

$$L_{Source} = E_{x \sim P_{data}}[\log p(s = real \mid x_{real})] + E_{z \sim P(z)}[\log p(s = generated \mid x_{generated})] \tag{5.3}$$

$$L_{Class} = E_{x \sim P_{data}}[\log p(Class = c \mid x_{real})] + E_{z \sim P(z)}[\log p(Class = c \mid x_{generated})] \tag{5.4}$$

For discriminator, training goal is to maximize both log-likelihood $L_{Source} + L_{Class}$, while generator is trained to maximize $L_{Class} - L_{Source}$.

The improved architecture ACGANs can generate data samples with better quality owing to additional class label information, so that it is suitable for supervised classification tasks.

Similarity Evaluation Criteria

The goal of using GAN architecture is to generate convincing data samples to realize data augmentation, therefore, the quality of generated data samples is significant. It is important to evaluate the similarity between generated data and the real data. Model evaluation for GAN architecture is still an open issue and several evaluation metrics are proposed to provide quantitative assessment, including inception score (IS) (Salimans et al. 2016), the Frechet inception distance (FID) (Heusel et al. 2017), and sliced Wasserstein distance (SWD) (Peyré and Cuturi 2019). For conventional image generation tasks, it is also possible to conduct visual evaluation. However, for time-series signals, especially mechanical sensor signals, it is not suitable to directly calculate the score using model pre-trained on natural images, and considering the inherent property of sensor data, time domain features and frequency distribution is able to reflect signal characteristics to some degree. Hence, the evaluation metrics is designed based on these statistical characteristics. In this chapter, we demonstrate two types of approaches to assess the performance of generative model: statistical indicators and experimental verification.

For the statistical features, Euclidean distance (ED), Pearson correlation coefficient (PCC), and Kullback–Leibler (K-L) divergence are introduced here to evaluate the similarity between generated samples and the real training data. These metrics are calculated to investigate the capability that generator model has for the distribution of training data.

ED is a direct way to measure the distance between two samples and evaluate their similarity. For a set of generated samples, the average ED is calculated to measure the distance between generated sample distribution and the real distribution. PCC is a measure of the correlation between two variables and is used to evaluate the strength of linear correlation between them. K-L divergence is a factor to evaluate the difference between two probability distributions.

Furthermore, a unique training and testing strategy is introduced to evaluate whether the generated samples are suitable as a training dataset independently in fault classification tasks. First, the real dataset is divided into two parts: training and testing datasets. Then, training dataset is used to train the generative model and a set of fake data is achieved. Then the dataset generated by the GAN is used as a training set to train a model, and the trained model is validated by testing data. The testing results including model loss and classification accuracy indicate the quality of the generated dataset and demonstrate the ability of the generated samples to be used for further practical applications.

GAN-BASED DATA AUGMENTATION FOR INTELLIGENT FAULT DIAGNOSIS

This section is going to introduce the framework of the GAN-based data augmentation for intelligent diagnosis and the model training procedure.

Framework for Intelligent Fault Diagnosis

This chapter aims to propose a framework utilizing ACGAN to generate high-quality artificial sensor signals for the purpose of data augmentation. The generated data serves to supplement imbalanced datasets in various applications. The overall architecture consists of two main components: data generation based on a GAN, and the assessment and application of the generated data as augmented data, as illustrated in Figure 5.2.

The first step involves designing an improved ACGAN architecture for generating sensor signals. The proposed ACGAN incorporates both the generator and discriminator. The generator generates samples from a latent space with specific labels and aims to fool the discriminator. The generated samples and real training data are combined and fed into the discriminator. The discriminator outputs two labels for each sample: one indicating whether the sample is real (output 1) or generated (output 0), and the other corresponding to a specific category.

Once the samples are generated, the dataset is augmented using the generated samples, denoted as $X_{generated}$. The next phase involves evaluating the quality of the generated data and applying both the generated and real data to real-world problems. Statistical characteristics and an experimental verification strategy are employed to assess the effectiveness of the generated data. The generated data is used to train a model, while the test data is utilized for validation. The classification accuracy is calculated to measure the similarity between the generated samples and real data.

In detail, 1D convolutional operations are used to construct the GAN. Convolutional operations have been proven to effectively learn local features, and an architecture

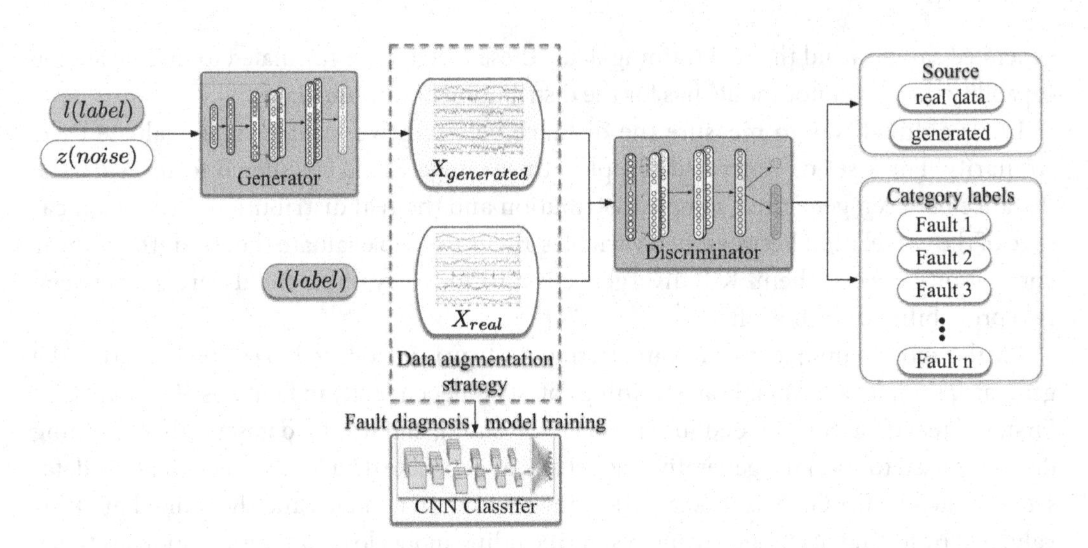

FIGURE 5.2 Data augmentation strategy using ACGAN for fault diagnosis task.

comprising several convolutional layers enables the learning of hierarchical representations. Unlike unsupervised learning, class conditional neural networks incorporate auxiliary information in the form of category labels. This inclusion aids model convergence and helps prevent model collapse. Batch normalization is added after the 1D-CNN layer to address the issue of gradient vanishing during the training process. Figure 5.3 provides detailed information about the proposed generator architecture.

In generator, the input layer is merged from noise input and class input, and it contains two up-sampling layers with size of 2. There are two layers of 1D-convolution operation followed by a batch normalization respectively with momentum 0.8. The first 1D-convolution has 16 feature maps with rectified linear unit (ReLU) as activation function and the kernel size is 16, while the second 1D-conlution layer has only 1 feature map with kernel size 16, using hyperbolic tangent as the activation function. The output of generator is 1D data sample.

As for the discriminator, the input layer is followed by one 1D-convolution layer with 8 kernels using LeakyReLU as activation function, and the kernel size is 16. Another 1D-convolution layer with 16 kernels using LeakyReLU is added and followed by dropout with the probability of 0.5. Then the model layer is flattened and linked to one fully connected layer with 0.5 dropout. Finally, two label prediction layers are added as the output layers.

Generator is designed to generate data from latent vector z, which is sampled from a uniform distribution over the interval (−1,1), denoted as $U(-1, 1)$. The outputs of generator form the fake dataset $X_{generated}$, and then they are sent to the discriminator together with real signal dataset. The discriminator is also built on 1D-CNN and the output layer adopts Sigmoid function to predict sample source and adopts Softmax function to predict specific labels.

Model Training Procedure

The ADAM optimizer is employed for model training, with a learning rate of 0.0001 for the discriminator and 0.0002 for the generator. The training procedure within each epoch can be summarized in three steps, as depicted in Figure 5.4:

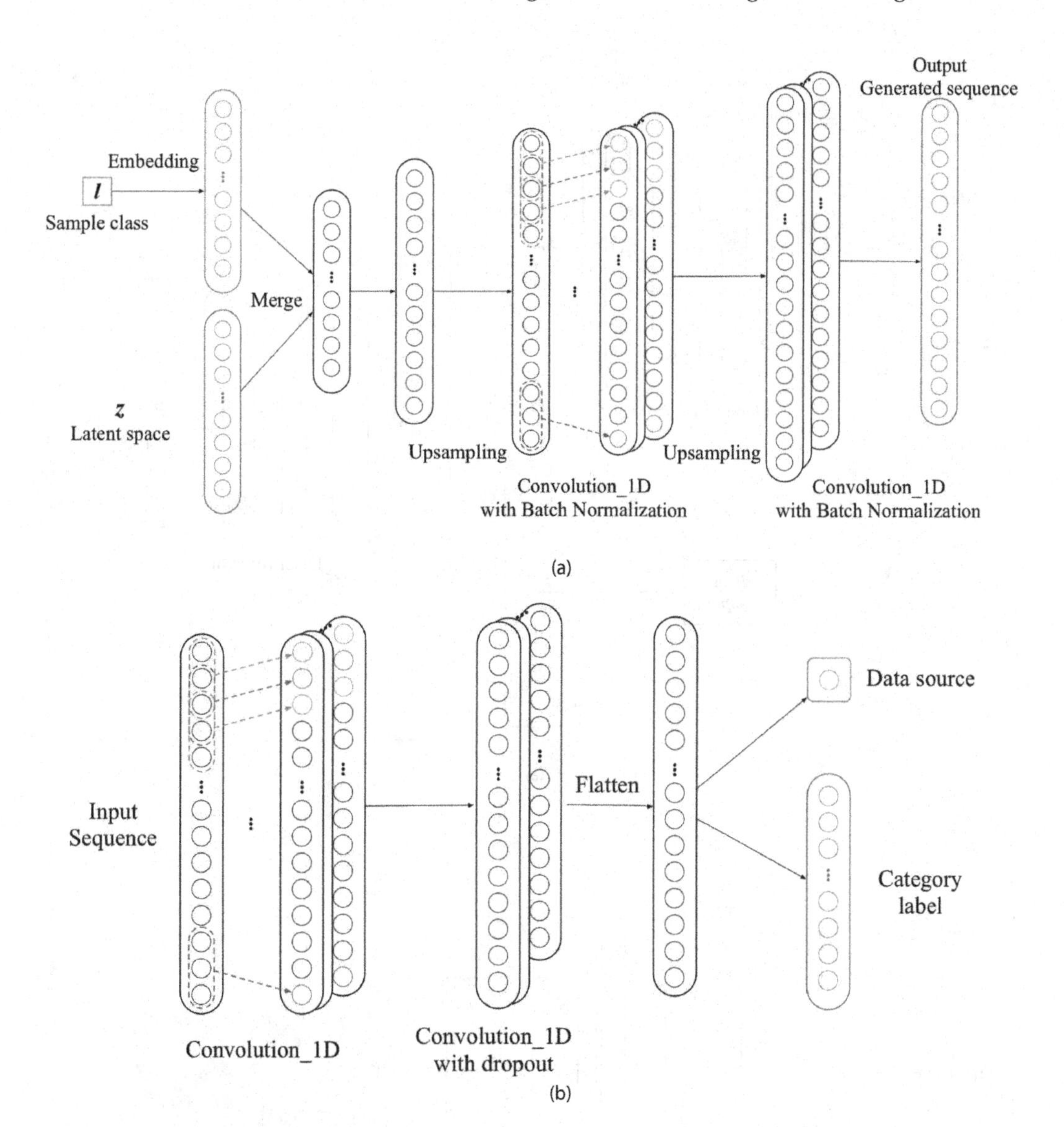

FIGURE 5.3 Architecture of the proposed (a) generator and (b) discriminator.

C1: The generator produces sequence samples from random noise of latent space with specific labels.

C2: Generated samples and real data samples are mixed together and sent to the discriminator. Based on loss function, discriminator is able to train using mixed data and labels and parameters in discriminator are able to be updated.

C3: After training the discriminator, the combined architecture starts to train. In this stage, the discriminator is set to be untrainable and parameters in it are frozen. During this step, only parameters in generator are able to be updated and the generator is trained to produce more realistic data samples. After training the combined structure, one epoch finished and training process goes on to start from ***C1*** again.

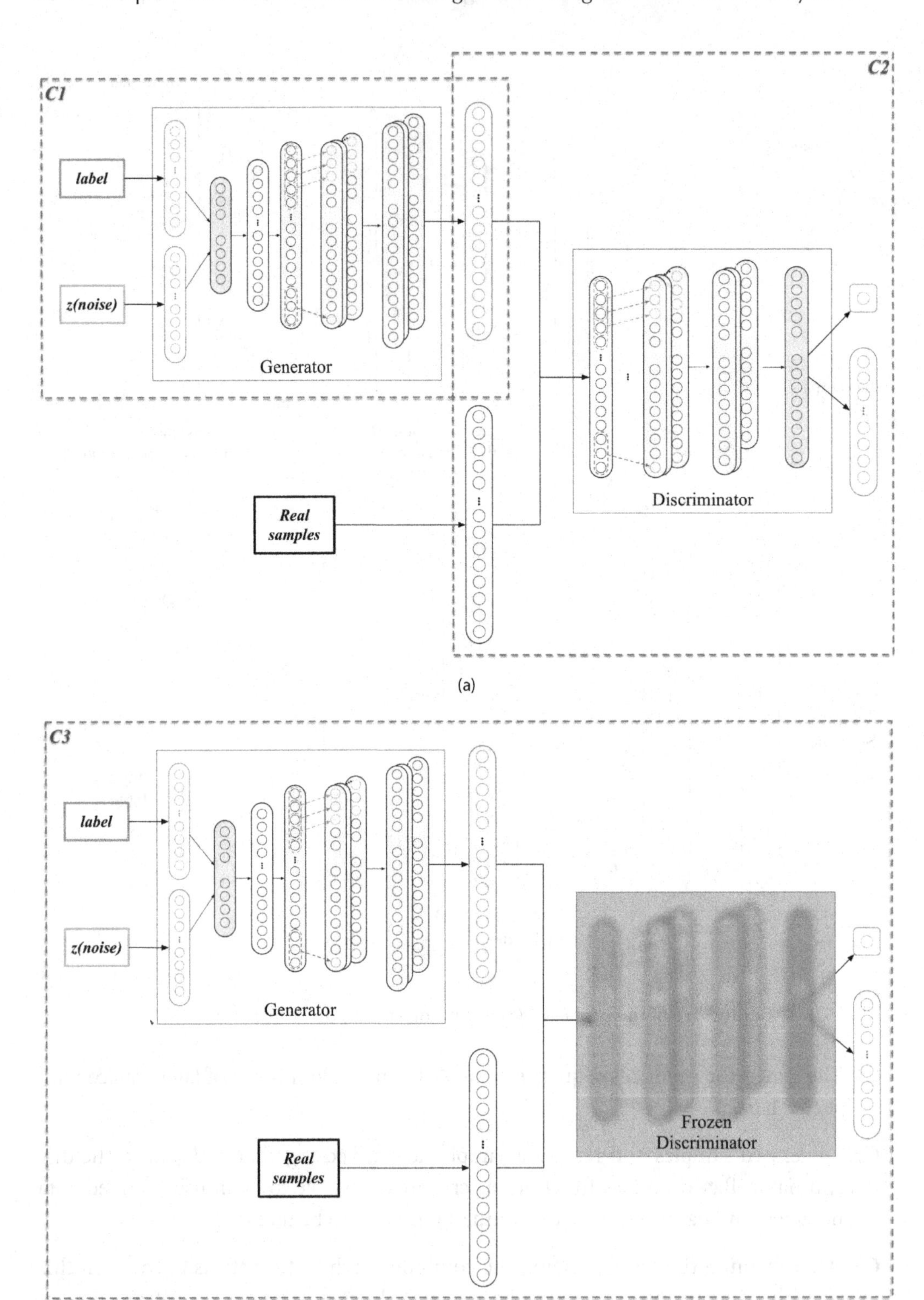

FIGURE 5.4 Adversarial training process of the proposed method.

Multiple training iterations have been carried out to force the whole model to achieve a balance between discriminator and generator. After enough iteration, the losses from generator and discriminator are able to achieve the balance, called Nash equilibrium, and generator can therefore produce realistic sensor data given certain labels.

EXPERIMENTAL STUDIES

In this section, the data from induction motor fault simulator is used for method evaluation.

Experimental Description

In order to investigate the performance of the proposed framework in practice, experimental verification is carried out among induction motor fault simulator (Yang, Yan, and Gao 2015), shown in Figure 5.5. An acceleration sensor is installed on the machine fault simulator and vibration signals are acquired during operation. Six different working conditions are simulated here and corresponding vibration signals are collected as the training dataset. Detailed information about each working state is listed in Table 5.1.

The sensor data are treated as the input of GAN, and a segment containing 4096 data points is regarded as one sample. Each working condition contains 300 separate samples and the whole dataset contains 1800 samples. The whole dataset is divided into two parts: training data and testing data. For each working condition, there are 200 time-series samples for training and 100 samples for testing. Training dataset is used for training the

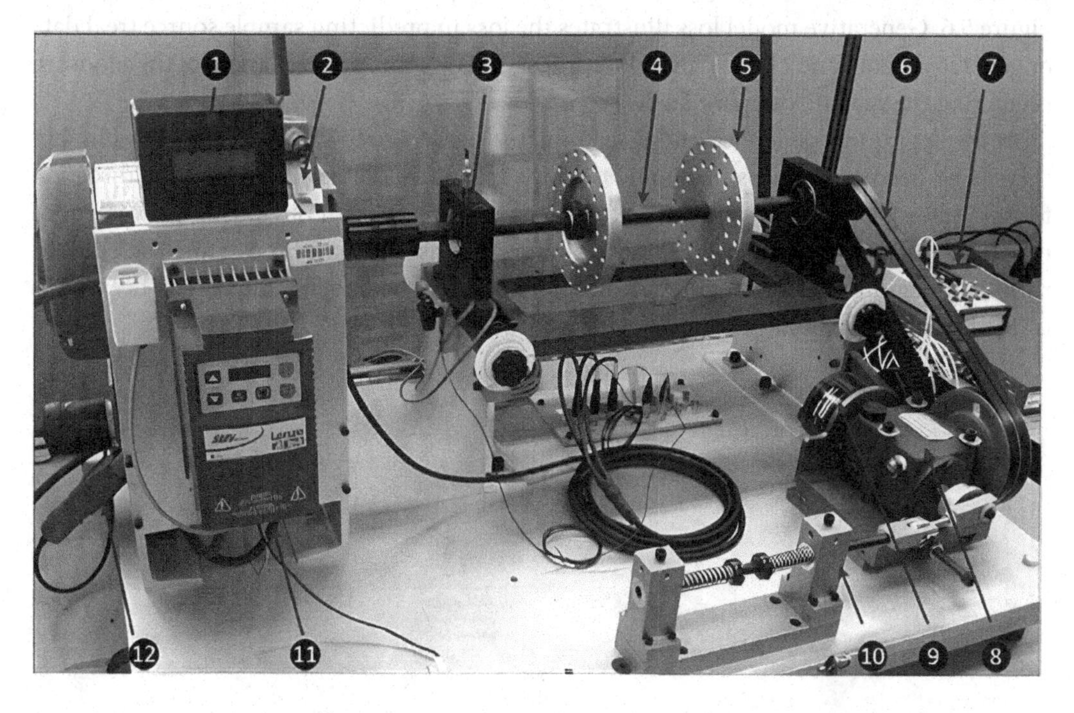

FIGURE 5.5 Experimental setup: (1) tachometer, (2) induction motor, (3) bearing, (4) shaft, (5) load disc, (6) belt, (7) data acquisition board, (8) bevel gearbox, (9) magnetic load, (10) reciprocating mechanism, (11) variable speed controller, and (12) current probe.

TABLE 5.1 Induction Motor Condition Descriptions

Condition		Description
HEA	Normal motor	Healthy, no defect
SSTM	Stator winding defect	3 turns shorted in stator winding
UBM	Unbalanced rotor	Unbalance caused by 3 added washers on the rotor
RMAM	Defective bearing	Inner race defect bearing in the shaft end
BRB	Broken bar	Three broken rotor bars
BRM	Bowed rotor	Rotor bent in center 0.01″

ACGAN architecture and generated samples are achieved, while testing dataset is only used in testing procedure to validate the ACGAN model and to evaluate the quality of generated samples. Testing dataset is not involved in any training process.

Results and Discussion

Model Training and Data Generation

In order to verify the generation ability of the proposed model, all of six kinds of sensor signals are generated. Training data together with corresponding labels are treated as the input of the discriminator in the proposed architecture, the generator samples from latent variable to model distribution of input data. Training epoch is set to be 100 and the number of generated samples for each working state is predefined as 1000. Experiments are carried out and achieve synthesized samples. Model performances are recorded, shown in Figure 5.6. Generative model loss illustrates the loss in predicting sample source (real data or generated data), and classification loss and accuracy show performances of the model in predicting the specific category labels.

At the beginning, generator and discriminator start to go toward Nash equilibrium but there is no decrease in classification loss which means model has not found the direction for optimal solution. After approximately 15 epochs, there is an obvious improvement in model classification accuracy and the generator and discriminator move to Nash equilibrium and begin to perform stably. After approximately 60 epochs, model has been well-trained and the losses stay around Nash equilibrium.

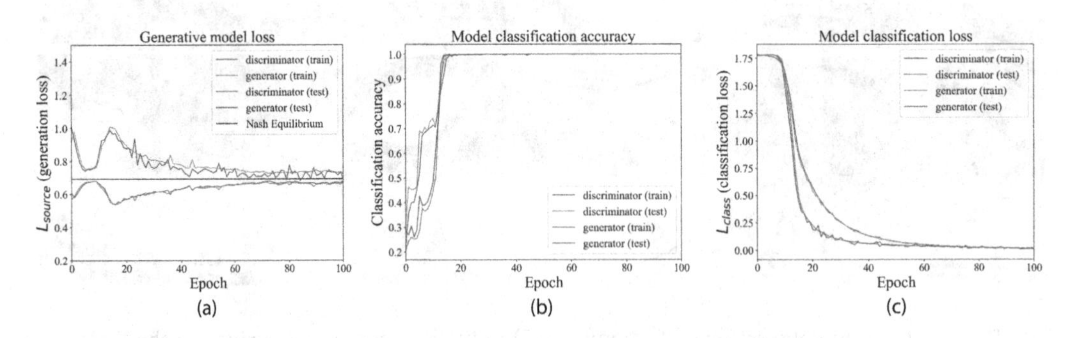

FIGURE 5.6 1D-CNN_BN model performance in (a) generative loss, (b) classification accuracy, and (c) classification loss.

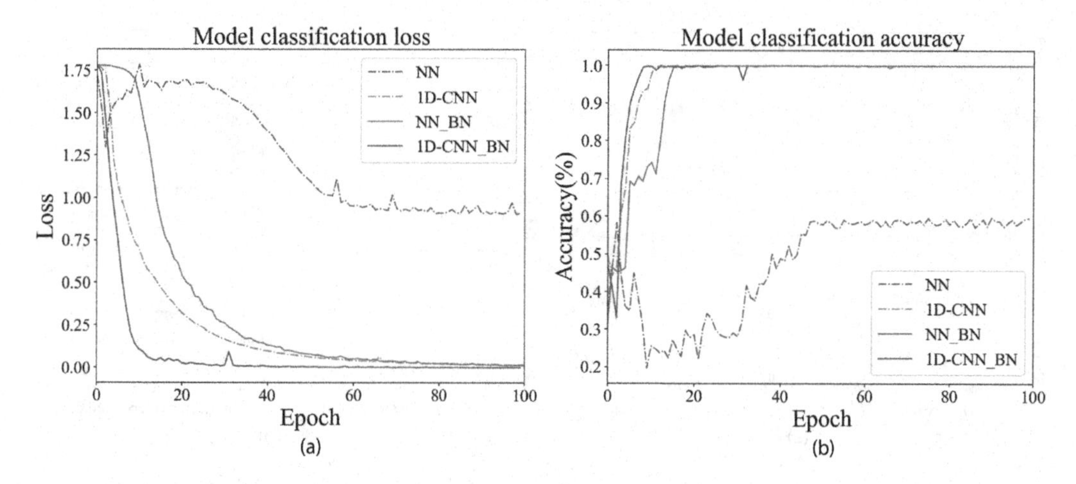

FIGURE 5.7 Model performances between various generator architecture (a) loss and (b) accuracy.

For comparison, we investigate several architectures in building generator, but the same discriminator built with 1D-CNN is used, including:

a. Generator built by three-layer fully connected neural network using backpropagation for generator training, denoted as NN.

b. Generator contains two layers of 1D-convolution operation without batch normalization, denoted as 1D-CNN.

c. Generator built by three layers fully connected neural network with batch normalization, denoted as NN_BN.

d. Our proposed framework, denoted as 1D-CNN_BN.

Hyper-parameters are set the same, and the model performances are illustrated in Figure 5.7 Model classification accuracy records the precision that the generative model provides the correct category labels.

From shown above, our proposed 1D-CNN_BN achieves the best performance in both classification accuracy and training speed. Generator built by three layers fully connected neural network is unable to accurately predict sample labels which means the model collapses and cannot generate reasonable samples. Neural network without batch normalization may be faced with the problem of overfitting, and the training procedure of the combined model may be easy to be dominated by the discriminator, where the generator model collapse. Neural network with batch normalization has much better performance than NN. Compared with 1D-CNN without batch normalization, our approach shows better model convergence rate. The experimental results show the effectiveness of our approach.

Data Generation Evaluation and Further Application

After training, generated samples from 1D-CNN_BN are achieved. Figure 5.8 shows both the time domain waveform and frequency spectrum of real sensor data and generated

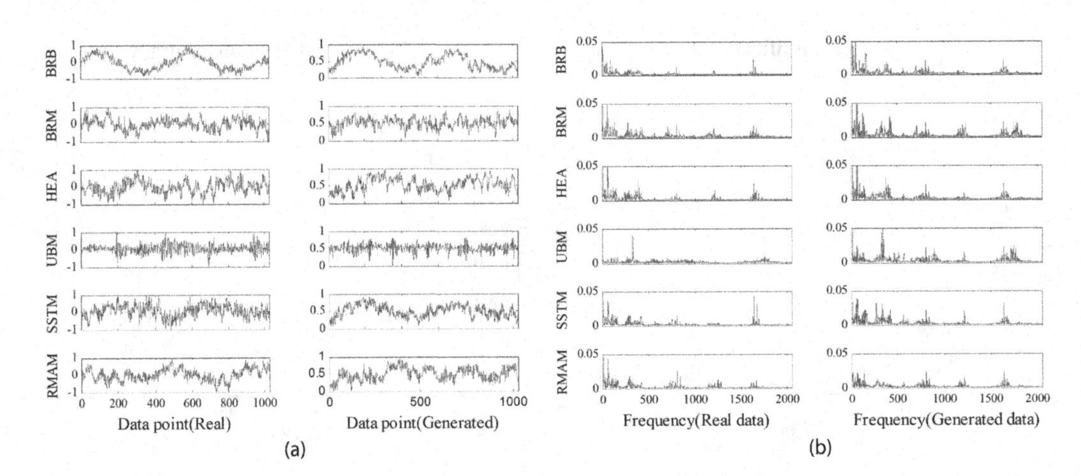

FIGURE 5.8 Time domain waveform (a) and frequency spectrum (b) of six different working conditions from real sensor data and generated samples.

samples among all six working conditions. From the figure, we can roughly recognize the similarity between real sensor data and generated samples.

Statistical indicators are calculated to evaluate the quality of the generated samples, shown in Table 5.2. ED indicates the distance between generated samples and the real data, and the smaller number indicates more similarity. In the same way, K-LD shows the divergence between two distributions, thus bigger number shows worse performance. To the contrary, PCC shows the correlation between the generated samples and the real ones, and a high correlation over 0.8 means strong similarity among these samples. Based on these metrics, our proposed approach gets better performances among various generator architecture and the samples generated from our method are alike the real data.

In order to further investigate the quality of generated samples, specific training strategy is carried out among generated datasets. The ACGAN model generates 6000 samples corresponding to 6 different working conditions, where each condition contains 1000 generated samples. These generated samples are then used to train a two-layer CNN model to realize fault classification. After fully training CNN model, testing data that are real sensor signals are sent to the model to predict possible labels. The classification accuracy using this strategy implies the similarity between generated model and real sensor signals. In this experiment, we obtain highly accurate classification accuracy 100% based on learning generated samples and testing in real data samples, which has shown that the generated samples are able to model the real data distribution.

TABLE 5.2 Statistical Features with Different Generator Architecture to Evaluate Generated Samples

Model	ED	PCC	K-LD
NN	0.3523	0.5899	0.3940
NN_BN	0.1487	0.8458	0.2011
1D-CNN	0.1204	0.8269	0.1581
1D-CNN_BN	0.0085	0.8664	0.1431

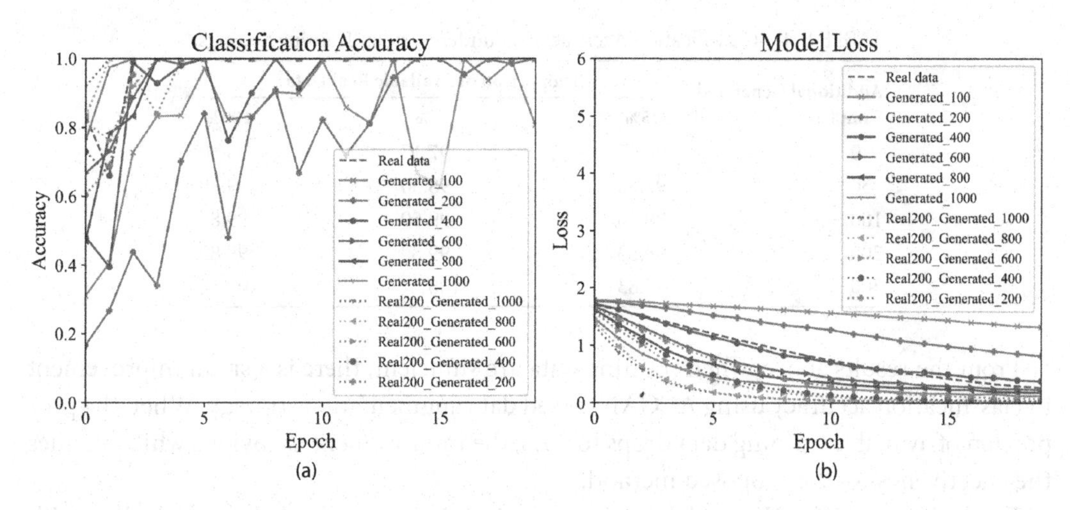

FIGURE 5.9 Fault diagnosis model performances among different training data settings: (a) accuracy and (b) loss.

To investigate the influence that the quantity of generated training samples has on model learning, we set various scenarios to test their performances. Classification accuracy and model loss are recorded during model learning, and performances on testing dataset in several selected scenarios are shown in Figure 5.9. Different scenarios have different sizes of real data samples and generated data samples, and the detailed information about training data settings of each scenario are listed in Table 5.3. The classification accuracy of these scenarios after enough training iterations are also shown in Table 5.3.

Based on the results shown above, it is obvious that training data with larger scale has better performance in classification accuracy and model converge speed. When using mixed data including real data and generated data, model has better performances than only using generated data. For enough training data, model using generated data is able to achieve highly accurate predictions.

Specifically, in order to investigate the effectiveness of ACGAN-based data augmentation strategy in helping improve fault diagnosis performances when the training data are limited, several experiments are designed. We simulate several circumstances where training data are limited in the form of various proportions of the available training data. In the meantime, different numbers of generated samples are used to give a comparison, as well as training without data augmentation. Classification results are shown in Table 5.4 under different settings.

TABLE 5.3 Different Training Data Settings Using Real Data and Generated Data and the Classification Results

	A	B	C	D	E	F	G	H	I	J	K	L
Real data (samples)	200	0	0	0	0	0	0	200	200	200	200	200
Generated data (samples)	0	100	200	400	600	800	1000	200	400	600	800	1000
Classification Accuracy (%)	99.80	95.33	99.50	99.83	99.83	100	100	99.83	99.83	100	100	100

TABLE 5.4 Classification Accuracy (%) under Various Training Settings

Additional Generated Samples	Proportion of Available Real Data		
	25%	50%	100%
0	95.67	97.17	99.80
50	97.50	98.50	99.80
100	98.50	99.50	99.83
200	98.83	99.80	99.83
300	98.83	100	100

From the results above, when training data are sufficient, there is a small improvement in classification accuracy using ACGAN-based data augmentation strategy. When the proportion of available training data drops to 25%, the improvement is obvious which verifies the effectiveness of the proposed method.

Furthermore, ACGAN-based data augmentation strategy is also helpful in dealing with unbalanced training data. Generated samples can be regarded as supplement samples to imbalanced training data, and the samples from minor category can be expanded by synthetic data. Finally, balanced training dataset is formed using real data and generated data. In order to investigate the performance of ACGAN-based model in unbalanced training dataset, we design several unbalanced datasets by reducing the number of training samples from one or more categories by 50%, and then ACGAN-based model is used to help generate balanced dataset. At last, both unbalanced original dataset and generated balanced dataset are utilized to train a two-layer CNN separately, and testing data are used to investigate the model performances. In this study, the induction motor dataset contains six different working conditions which means the dataset has six categories, so we design five different unbalanced datasets which have 1, 2, 3, 4, 5 classes of reduced samples, respectively. The minor class has only 100 samples while other classes contain 200 samples. In order to achieve balanced dataset, part of the unbalanced data is chosen to train the ACGAN model and generated samples for minor class are obtained. Then, samples of minor class are expanded to 200 by adding generated samples, where the synthetic dataset is balanced. Table 5.5 shows the various imbalanced datasets and their classification accuracies.

With unbalanced training dataset, deep model is not able to well learn the data distribution among input data. Using data augmentation is an alternative to supplement the minor classes so that the training dataset will be balanced among various classes. Classification results also verify the effectiveness among different unbalanced settings.

TABLE 5.5 Classification Accuracy under Different Unbalanced Dataset and Generated Dataset

Number of Classes with Reduced Samples	Unbalanced Dataset	Generated Balanced Dataset
1	83.33%	99.83%
2	66.67%	99.50%
3	50.00%	99.33%
4	75.67%	99.33%
5	94.50%	99.16%

In addition, ACGAN-based data augmentation strategy starts an open issue in generating raw data samples in machine health monitoring. Investigation on building generative models and generating reasonable samples helps better understand sensor signal distribution and develop reliable predictive system in fault diagnosis and prognosis.

CONCLUSION

This chapter introduces a framework based on ACGAN for generating artificial raw data of mechanical sensor signals. The primary objective of this framework is to address the issue of unbalanced data in training deep architectures by utilizing high-quality generated samples for data augmentation. To assess the quality of the generated data, statistical characteristics are computed as a benchmark for evaluating the disparity between synthesized data and real sensor signals. In addition, a testing strategy is presented to validate the effectiveness of the generated samples. This involves training a classifier using the generated samples and evaluating its performance on real data. The results of the classification experiments demonstrate that the proposed framework is capable of producing convincing sensor data. The ACGAN-based method proves to be a valuable data augmentation technique for handling unbalanced datasets and contributes to subsequent classification tasks.

REFERENCES

Arjovsky, Martin, Soumith Chintala, and Léon Bottou. 2017. "Wasserstein GAN." *arXiv preprint arXiv: 1701.07875.* https://doi.org/10.48550/arXiv.1701.07875.

Donahue, Chris, Julian McAuley, and Miller Puckette. 2018. "Synthesizing audio with generative adversarial networks." *arXiv preprint arXiv: 1802.04208.* https://doi.org/10.48550/arXiv.1802.04208.

Goodfellow, Ian J., Jean Pouget-Abadie, Mehdi Mirza, Bing Xu, David Warde-Farley, Sherjil Ozair, Aaron Courville, and Yoshua Bengio. 2014. "Generative Adversarial Networks." *Advances in Neural Information Processing Systems.* https://doi.org/10.48550/arXiv.1406.2661.

Gulrajani, Ishaan, Faruk Ahmed, Martin Arjovsky, Vincent Dumoulin, and Aaron Courville. 2017. "Improved Training of Wasserstein GANs." *Advances in Neural Information Processing Systems.* https://doi.org/10.48550/arXiv.1704.00028.

Hartmann, Kay Gregor, Robin Tibor Schirrmeister, and Tonio Ball. 2018. "EEG-GAN: Generative Adversarial Networks for Electroencephalograhic (EEG) Brain Signals." *arXiv preprint arXiv: 1806.01875.*

Heusel, Martin, Hubert Ramsauer, Thomas Unterthiner, Bernhard Nessler, and Sepp Hochreiter. 2017. "GANs Trained by a Two Time-Scale Update Rule Converge to a Local Nash Equilibrium." *Advances in Neural Information Processing Systems* 30. https://doi.org/10.48550/arXiv.1706.08500.

Odena, Augustus. 2016. "Semi-Supervised Learning With Generative Adversarial Networks." *arXiv preprint arXiv: 1606.01583.* https://doi.org/10.48550/arXiv.1606.01583.

Odena, Augustus, Christopher Olah, and Jonathon Shlens. 2016. "Conditional Image Synthesis With Auxiliary Classifier GANs." *arXiv preprint arXiv: 1610.09585.* https://doi.org/10.48550/arXiv.1610.09585.

Peyré, Gabriel, and Marco Cuturi. 2019. "Computational Optimal Transport." *Foundations and Trends in Machine Learning* 11 (5-6): 355–607. https://doi.org/10.48550/arXiv.1803.00567.

Radford, Alec, Luke Metz, and Soumith Chintala. 2016. "Unsupervised Representation Learning With Deep Convolutional Generative Adversarial Networks." *arXiv preprint arXiv: 1511.06434.* https://doi.org/10.48550/arXiv.1511.06434.

Salimans, Tim, Ian Goodfellow, Wojciech Zaremba, Vicki Cheung, Alec Radford, and Xi Chen. 2016. "Improved Techniques for Training GANs." *arXiv preprint arXiv: 1606.03498*. https://doi.org/10.48550/arXiv.1606.03498.

Yang, Xueliang, Ruqiang Yan, and Robert X. Gao. 2015. "Induction Motor Fault Diagnosis Using Multiple Class Feature Selection." Proceedings of the IEEE International Instrumentation and Measurement Technology Conference (I2MTC), 256–260. https://doi.org/10.1109/I2MTC.2015.7151275.

CHAPTER 6

Multi-Sensor Fusion for Intelligent Fault Diagnosis

INTRODUCTION

Nowadays, induction motor is widely used in mechanical systems and is an essential component of manufacturing machines where incipient faults during machine operation can pose threats to system reliability. The failures of induction motor in operation may cause accidents, which can bring huge economic losses and even casualties. Therefore, it is important to evaluate the working state of induction motors, detect potential faults, and then take proper action to prevent major machine failure from happening (Elbouchikhi et al. 2017). Accordingly, induction motor fault diagnosis plays a significant role in safety production and equipment maintenance, and it can improve machine quality, enhance working safety, and reduce maintenance cost (Sun et al. 2017).

Many research studies have been carried out for developing reliable and efficient induction motor fault diagnosis techniques (Hassan et al. 2018). For decades, many fault diagnosis methods have been based on motor current analysis which is called motor current signature analysis (MCSA). After data acquisition, critical features associated with the working state are extracted from motor current data using signal processing techniques. MCSA has been one of the most widely used fault diagnosis techniques for induction motor as it is inexpensive and noninvasive (Tian et al. 2018), and it turns out to be successful in detecting stator faults and rotor faults (Da Silva, Povinelli, and Demerdash 2008). Sapena-Bano et al. (2015) proposed a framework using fast Fourier transform (FFT)-based stator current analysis to diagnose the faults of electrical machines in transient conditions. Spectral analysis of the stator current based on Hilbert transform was also utilized (Noureddine, Hafaifa, and Kouzou 2017). Although MCSA has achieved much success in various applications, sometimes its performance is influenced by voltage imbalance, induced harmonics, and other disturbances. Recently, vibration signals and acoustic signals have also been utilized to analyze induction motor working conditions, especially in detecting bearing faults (Delgado-Arredondo et al. 2017). By recognizing

DOI: 10.1201/9781003474463-8

features extracted from acoustic signals, early fault detection of the single-phase induction motor can be achieved (Glowacz and Glowacz 2017).

However, to achieve reliable and robust induction motor fault diagnosis and realize the detection of several faults simultaneously, including stator faults, rotor faults, and bearing faults, various types of sensor signals should be leveraged together. This is because each type of sensor signal contains specific information related to certain working condition. Thus, combining these signals and analyzing them together can be an effective way to realize multi-fault diagnosis under complex conditions.

In fact, feature extraction from sensor signals is a crucial step, and the performance of the fault diagnosis method depends on how precise the extracted features can represent the fault signature of the monitored induction motor. Nevertheless, the performance of conventional feature-based fault diagnosis depends on the quality of extracted features and its suitability to a certain task. This requires expert knowledge and human intervention, consequently bringing uncertainty to overall performance. Compared with traditional ways in fault diagnosis, deep learning (DL) method can learn representations and patterns hierarchically from the input data and select those that can best represent machine working condition through gradually adjusting connected weights. This reduces the influence caused by human intervention and improves efficiency. By making use of the deep architecture and massive training data, DL method has the ability to model complex working conditions and output accurate predictions. There are several typical deep architectures that have been successfully applied to machine fault diagnosis, including autoencoders (AEs) (Vincent et al. 2008), deep belief networks (DBNs) (Hinton, Osindero, and Teh 2006), and convolutional neural networks (CNNs) (Rawat and Wang 2017). For example, an intelligent approach based on stacked denoising AE and feature ensemble was applied to motor fault diagnosis (Wang et al. 2017). Furthermore, one-layer AE-based neural network was proven to be effective in the task of fault classification for induction motors (Sun et al. 2016). DBN model was successfully applied for induction motor fault diagnosis using the frequency spectrum of vibration signals (Shao et al. 2017). A DBN-based classification model was proposed and the model was verified in aircraft engine application (Tamilselvan and Wang 2013). One-dimensional CNN was built to analyze the raw time-series data of motors and proved to be successful in diagnosing fault states (Ince et al. 2016), and a new CNN architecture based on LeNet-5 was proposed to process motor bearing dataset (Wen et al. 2017). It can be seen that all the previous researches utilize one single type of signal. However, it can only characterize partial information of the induction motor. As the information related to different fault signatures in a single signal is limited, there is a limitation in fault classification accuracy in real applications.

Aiming at overcoming the limitation caused by single-sensor signal and improving fault classification accuracy and model stability, a multi-signal CNN-based approach using time–frequency distribution (TFD) image is proposed in this chapter, and both the vibration and current signals are leveraged. Compared with previous works, the proposed deep architecture is able to learn from multiple input signals simultaneously so that the classification accuracy and the stability of the deep model are improved.

Through model training, the proposed deep architecture is able to automatically learn and select suitable features that contribute to accurate classification, which reduces overall human intervention.

The rest of this chapter is organized as follows: background of multi-sensor fusion, the proposed deep CNN-based method for induction motor fault diagnosis, experimental studies, and conclusion.

BASIC METHODS FOR MULTI-SENSOR FUSION

This section is going to introduce basic methods for multi-sensor fusion. The concept of multi-sensor fusion in fault diagnosis can generally be categorized into three strategic approaches: data-level fusion, feature-level fusion, and decision-level fusion.

Data-Level Fusion

Data-level fusion is aimed at the data processing stage of the fault diagnosis framework, which aligns measuring data from multi sensors to form a unified and integrated sensor data. The integrated data be considered as the primary signal for further processing. Data-level fusion generally adopts a centralized fusion system for fusion processing, which is the lowest level of fusion. Among the three fusion schemes, it possesses the minimum data loss and the highest reliability. This method makes higher requirements on the computing power of the hardware platform, so the current applications focus on scientific research.

Feature-Level Fusion

Feature-level fusion is a fusion of features oriented toward monitoring objects, which is usually performed after obtaining raw data and extracting features. This method extracts the feature vectors contained in the collected data to reflect the attributes of the monitoring objects. After convolution and pooling of the fused feature information from sensors, a dataset with sensor feature information will be obtained. Finally, by annotating and training the dataset, the diagnostic model can be acquired.

Decision-Level Fusion

Decision level fusion employs the data features obtained through feature level fusion to perform certain discrimination, classification, and simple logical operations. Advanced decisions are made based on application requirements, resulting in application-oriented fusion. Due to the need to provide a separate model for each modality, it can better model different modal data features, and can cope with partial data loss and asynchrony between certain modalities. However, it loses data correlation at the feature level, which usually has higher implementation difficulty.

DCNN-BASED MULTI-SENSOR FUSION FOR INTELLIGENT FAULT DIAGNOSIS

This section is going to introduce TFD and DCNN-based multi-sensor fusion for intelligent fault diagnosis.

Time–Frequency Distribution

TFD is used to describe the relationship between a signal's information in the time domain and its corresponding spectral content in the frequency domain (Cohen 1989). It is an effective approach in feature extraction as it reserves and reveals both time and frequency information of the original signal simultaneously. There are a variety of methods that are able to generate time–frequency representation from sensor signals. Due to the property of good time–frequency resolution, wavelet transform is widely adopted in analyzing sensor signals measured on induction motors.

Continuous wavelet transform (CWT) is widely used and can be calculated by the inner product of the signal and a family of the wavelets (Lin and Qu 2000). Scaling and translating the mother wavelet $\Psi(t)$ generates the wavelet family:

$$\psi_{s,\tau} = \frac{1}{\sqrt{s}}\psi\left(\frac{t-\tau}{s}\right) \tag{6.1}$$

where s and τ denote the scale parameter and the translation parameter, separately, and s is related to frequency inversely.

The CWT of given signal $x(t)$ can be obtained by the convolution operation of the complex conjugate, which is mathematically defined as follows:

$$\psi_{s,\tau} = \frac{1}{\sqrt{s}}\psi\left(\frac{t-\tau}{s}\right) W(s,\tau) = \langle x(t), \psi_{s,\tau}\rangle = \frac{1}{\sqrt{s}}\int x(t)\psi^{*}\left(\frac{t-\tau}{s}\right)dt \tag{6.2}$$

where $\psi^{*}(\cdot)$ represents the complex conjugate of function $\psi(\cdot)$. Through that, the signal is decomposed into a series of wavelet coefficients. As there are two kinds of parameters, scale and translation parameters, the signal can be projected to a 2D plane. Therefore, time-scale (or frequency) images can be obtained by conducting CWT to the original signals.

DCNN-Based Multi-Sensor Fusion for Intelligent Fault Diagnosis

A fault diagnosis framework based on DCNN for detecting the working state of induction motors using multiple sensor signals has been developed. The proposed method can automatically classify induction motor condition by learning patterns which contain crucial information for distinguishing different working states directly from the original signals.

In general, multiple sensor signals are utilized for the purpose of fault diagnosis. In induction motor fault diagnosis tasks, vibration signal and current signal are commonly used as they both contain useful fault-related signature reflecting the certain working state. Therefore, both vibration signal and current signal are used in this chapter, and their TFD images that provide inherent discriminative information among different working conditions are used as the input. Figure 6.1 shows examples of original vibration and current signals collected from the induction motor under three different fault conditions.

As shown in Figure 6.1, the original signals in time domain do not have clear discriminative features and it is difficult to correctly distinguish different working states. Therefore, data preprocessing using CWT is implemented as it generates time-frequency images

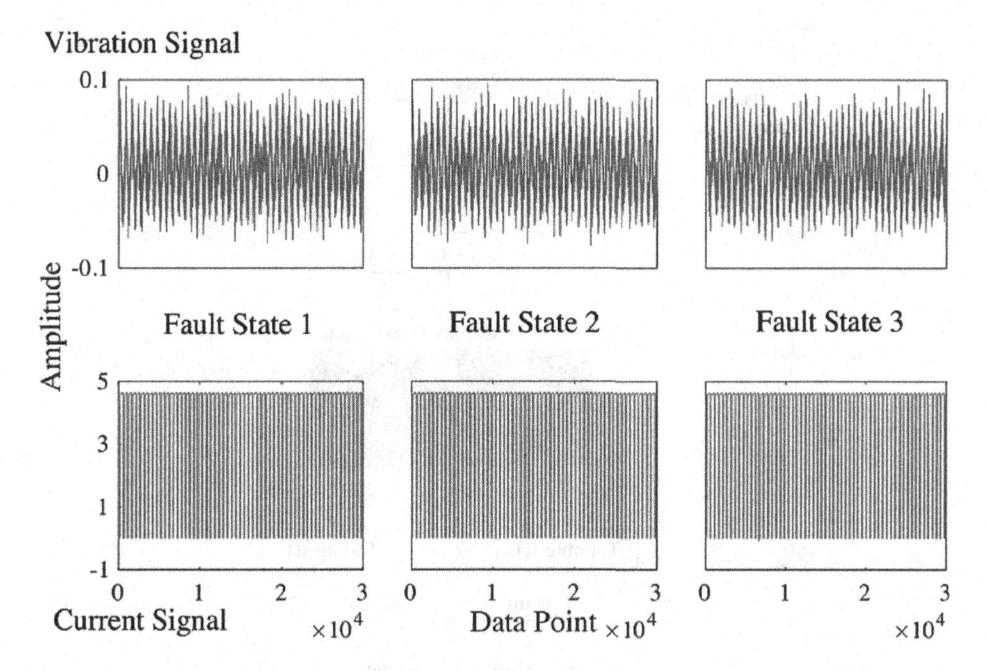

FIGURE 6.1 Original vibration signals and current signals from induction motor under three different fault states.

containing fault signature among different conditions where DL method can be used to learn the certain hidden pattern from the images, and examples are shown in Figure 6.2.

Figure 6.3 illustrates the flowchart of the presented DCNN-based multi-signal induction motor fault diagnosis, where both vibration signal and current signal are simultaneously measured by sensors attached on the induction motors under operation, and then these signals are converted to TFDs using CWT. These obtained TFDs are treated as gray-scale images and

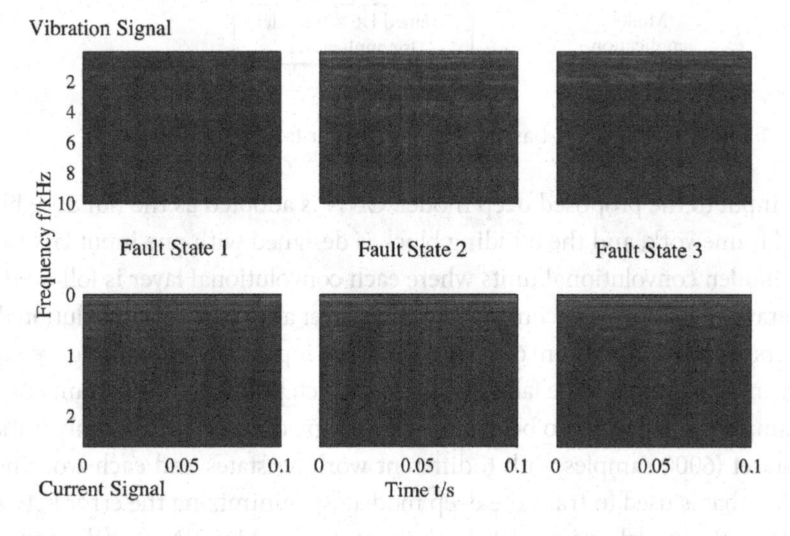

FIGURE 6.2 TFDs of vibration signals and current signals from induction motor under three different fault states.

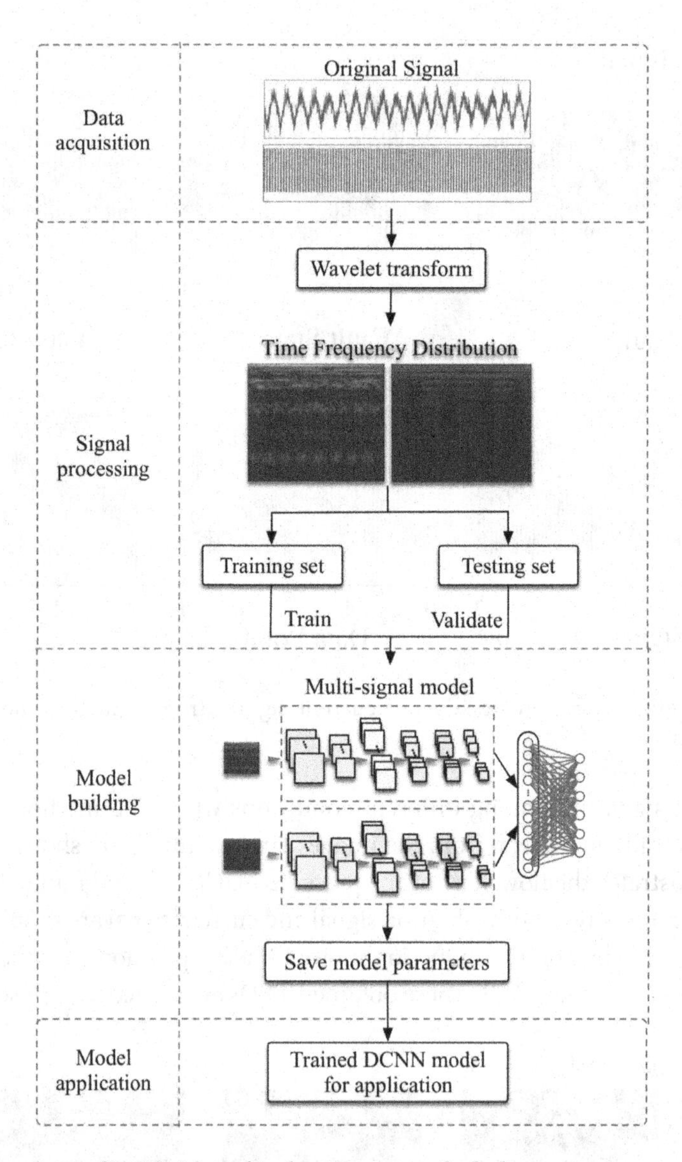

FIGURE 6.3 Flowchart of DCNN-based induction motor fault diagnosis.

used as the input to the proposed deep model. CNN is adopted as the building block of our multi-signal framework, and the building block is designed with one input layer, one output layer, three hidden convolutional units where each convolutional layer is followed by a max-pooling operation and one fully connected hidden layer after pairs of convolutional and max-pooling layers, as shown in Figure 6.4. The size of the input layer equals to gray-scale images size, and the model output is the label prediction which points to one certain condition that the input sample is most likely to belong to. The designed model is randomly initialized and training dataset (6000 samples with 6 different working states and each working state has 1000 samples) that is used to train the deep model. By minimizing the error between output prediction from the model and real labels, the parameters of DCNN model are updated iteratively. After enough training epochs, the model is well trained and all fine-tuned parameters are saved, which can then be applied to diagnosing induction motor working conditions.

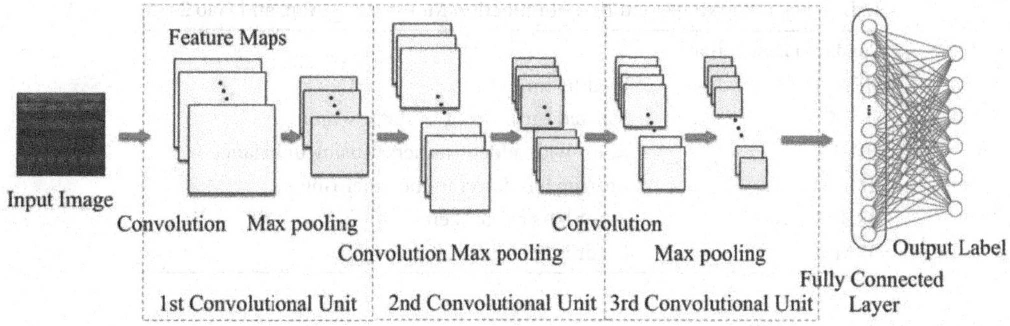

FIGURE 6.4 DCNN as a building block of multi-signal architecture.

EXPERIMENTAL STUDIES

Experimental Description

Data Preprocessing and Splitting

To verify the performance of the presented approach, the sensor data acquired on a machine fault simulator are used in this chapter. Vibration signals and current signals under six different working conditions are measured simultaneously, as illustrated in Figure 6.5. The description on each working condition is listed in Table 6.1.

In order to prepare dataset for training and testing, the acquired signals are first preprocessed to corresponding time–frequency images through CWT. Specifically, for both vibration

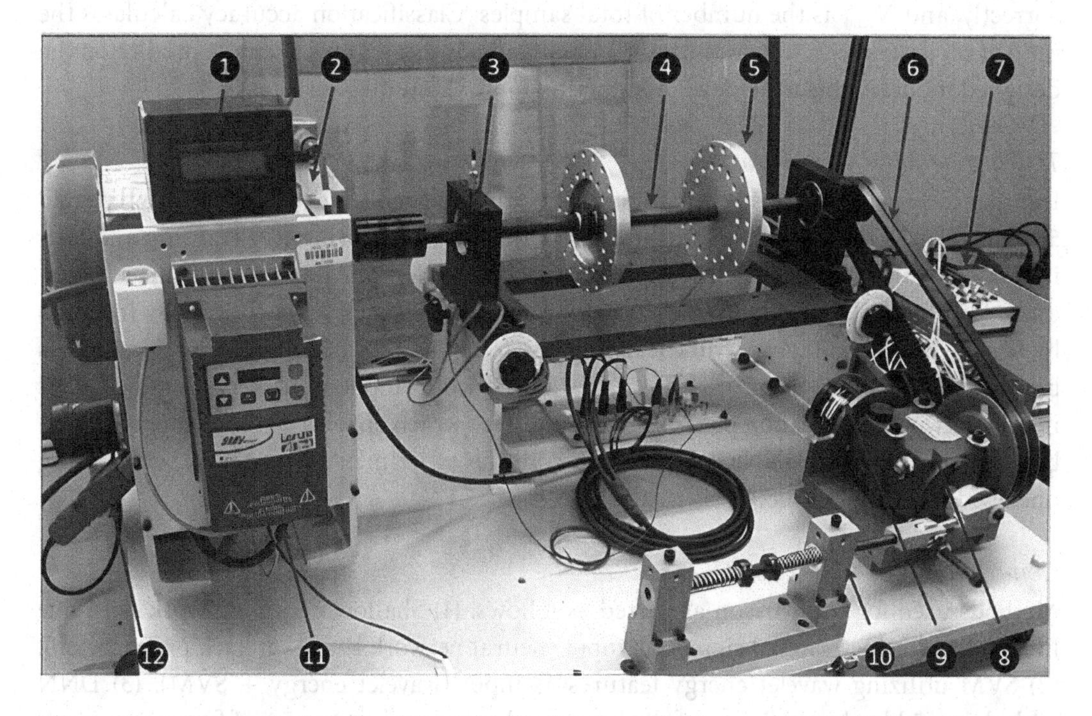

FIGURE 6.5 Experimental facility (Yang, Yan, and Gao 2015). (1) Opera meter, (2) induction motor, (3) bearing, (4) shaft, (5) loading disk, (6) driving belt, (7) data acquisition board, (8) bevel gearbox, (9) magnetic load, (10) reciprocating mechanism, (11) variable speed controller, and (12) current probe.

TABLE 6.1 Working Condition of Induction Motor (Yang, Yan, and Gao 2015)

Condition Description	
HEA	Healthy motor
SSTM	Shorted turns occur in stator winding
UBM	Rotor with added washers causing unbalance
RMAM	Bearing has defect in the inner ring
BRB	Rotor bars are broken
BRM	Rotor is bent

and current signals, 1024 data points are chosen to be one sample and time–frequency image of each sample is reshaped to be 128 × 128 pixels. In all, 1000 samples for each working condition are acquired on the machine fault simulator and randomly divided into training and testing subsets, with 800 training samples and 200 testing samples. As 6 different working conditions are investigated, the overall dataset contains 6000 samples.

Evaluation Metrics

Classification accuracy is calculated as the metrics to evaluate the performances of the deep architectures:

$$\text{Accuracy} = \frac{n_{\text{right}}}{N_{\text{total}}} \tag{6.3}$$

where n_{right} represents the number of samples that the deep architecture is able to classify correctly, and N_{total} is the number of total samples. Classification accuracy calculates the ratio of accurate prediction to the total dataset and it indicates how accurate prediction the designed model is able to achieve in induction motor fault diagnosis tasks.

Training Details

In order to provide a robust estimation of the proposed model, experiments are repeated 50 times and statistics are used to summarize the model performance. Training and testing time are also recorded to investigate the model efficiency. All the algorithms are conducted in Keras platform using TensorFlow backend. TensorFlow was originally invented by Google Brain team for machine learning researches and for training DNNs. Now, it has become an open-source software library for numerical computation. Keras is a high-level neural networks application programming interface which enables fast implementation in building and training DNNs. We conducted all the experiments on a Linux server using one Nvidia GTX GPU 1080nosis tasks.

Compared Models

Methods used for comparison are listed as follows: (1) Shallow neural network with one hidden layer using original signals as input (neural network based on back propagation). (2) SVM utilizing wavelet energy features as input (wavelet energy + SVM). (3) DNN with three hidden layers that are fully connected using wavelet energy of features as input (wavelet energy + DNN). (4) One-layer SAE-based neural network using vibration signal only (SAE). (5) MCFS using selected features. (6) 1D CNN-based discriminative feature learning approach using vibration signal only (CDFL). (7) DBN with four hidden layers using original signals as input (DBN).

Results and Discussion

Single-Signal Fault Diagnosis

For comparison, we conduct single-signal fault diagnosis using vibration signal and current signal separately. The image dataset that contains six working conditions is used as the input of single-signal DCNN. The built network outputs predict label that corresponds to one working condition. For vibration signals, the network contains three 2D convolutional layers with kernel sizes 6, 8, and 6, and filter numbers 6, 12, and 12, respectively, where each convolutional layer is connected to a max-pooling layer which sets the pooling size as 2 × 2 and the stride length as 2. The activation function used in the convolutional layer is ReLU. Afterward, the learned 2D representation is flattened and fully connected to a hidden layer with 512 ReLUs. Finally, the output layer using the Softmax activation function with six units corresponding to six different working conditions can predict the label of each input image. For current signals, the built network has three 2D convolutional layers with filter numbers 6, 12, and 12, respectively, activated by ReLU. The kernel sizes of filters are 3, 3, and 6, respectively. Each convolutional layer is followed by a max-pooling layer where the pooling size is set to be 2 × 2, and the stride length is set as 2. The fully connected layers have 512 ReLUs and the output layer is the same as the one for dealing with vibration signals. The model weights of both architectures are randomly initialized, and the cross-entropy error is calculated to update the model parameters using Adam optimizer with learning rate being set to be 0.0001. The batch size is 32 and the whole training process is finished when the classification accuracy stops improving.

Experimental results of vibration and current signals are listed in Table 6.2, model loss and accuracy during training and testing are shown in Figure 6.6, and the confusion matrix to clearly illustrate the classification rate of each category is shown in Figure 6.7.

It can be seen from the results that for training procedure, DCNN-based single-signal model gets better classification accuracy in current signals than that in vibration signals. It means that based on the existing dataset, it is relatively difficult to learn features from vibration signals, and the model is not well-trained using vibration signals. For testing dataset, the proposed model has a better performance in vibration dataset, which indicates the model has worse generalization capability using current signals. Although the proposed model gets faster convergence in current signals, it also shows a gap both in model loss and model accuracy between the training set and testing set which indicates

TABLE 6.2 Classification Results in Single-Signal Model

	Classification Accuracy Mean (Standard Deviation)		**Computation Time**	
	Train	**Test**	**Train**	**Test**
Vibration	99.16% (0.14)	98.72 (0.26)	201 s	0.6 s
Current	100% (0.00)	98.26 (4.96)	156 s	0.5 s

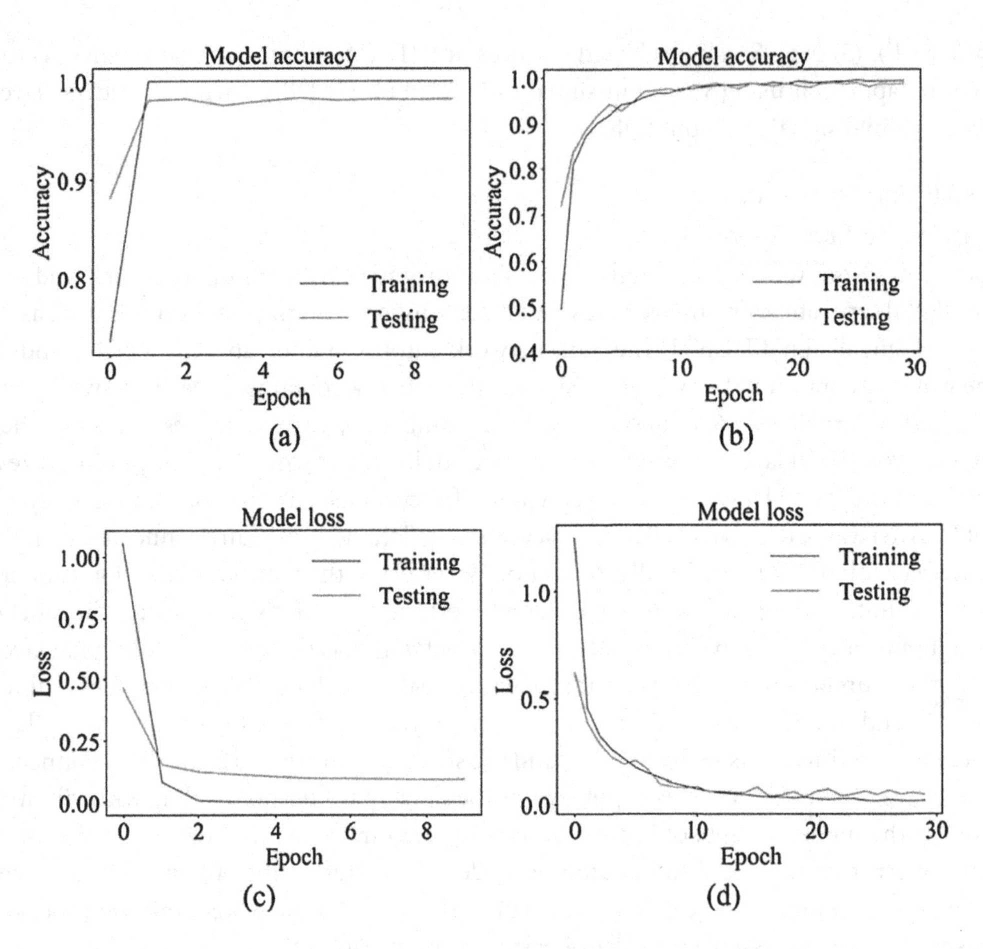

FIGURE 6.6 Model loss and classification accuracy for single-signal model. (a) Current accuracy, (b) vibration accuracy, (c) current loss, and (d) vibration loss.

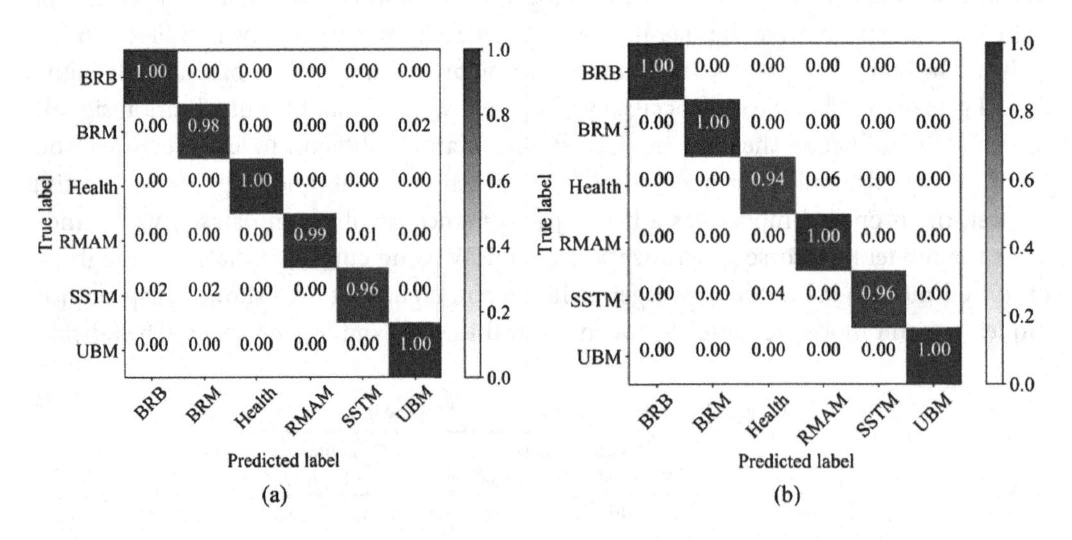

FIGURE 6.7 Confusion matrix for multi-signal model. (a) Multi-channel model and (b) merged model.

a slight overfitting problem in current signals. The standard deviation of model accuracy in 50 trails shows that DCNN model has better stability in vibration signals (0.26) in the testing set than current signals (4.96). Besides, from confusion matrix, DCNN model can achieve accurate classification in recognizing induction motor health state using vibration signals, but using current signals it only gets 94% accuracy. On the contrary, when recognizing fault states of BRM and RMAM, using current signals has much more accurate results than vibration signals.

These problems can be caused by limited training data and the massive amount of sensor data during industrial manufacturing is expensive to acquire. Therefore, combining multiple sensor signals together to enlarge the dataset and leveraging unique features of each signal simultaneously is a solution that contributes to a better performance in the learning model.

Multi-signal Fault Diagnosis

For multi-signal fault diagnosis, two different architectures are investigated to explore the optimal performance. One is to combine different signals to form a multi-channel input, which is used as the input of the DCNN that can learn features from various types of signals simultaneously, as shown in Figure 6.8(a). The other is to train separate DCNN without fully connected layers using each of the signals, and then these separate networks are merged together in higher layers to perform final classification, as shown in Figure 6.8(b). The same dataset is used in both architectures in order to compare their performances.

Architecture 1-Multi-channel Model: Multiple signals are combined in the data preprocessing procedure, where two different time–frequency images from vibration signals and current signals are joined together to form a two-channel image with each channel representing one type of signal. As a result, each sample has the size of $2 \times 128 \times 128$, and then the combined dataset is used as the input of the deep model which contains three 2D convolutional layers with kernel sizes 6, 8, and 6, and filter numbers 12, 24, and 24, respectively, followed by max-pooling operation with pooling size as 2×2, and stride length as 2. After flattened, the model is followed by a fully connected layer which has 1024 units. ReLUs are utilized as the activation function in both convolutional layers and fully connected layers. The output layer has six units corresponding to six different labels, which is the same as the one used in single-sensor signal model.

Architecture 2-Merged Model: Two separate convolutional networks are used to analyze different sensor signals, and the separate networks are merged in fully connected layers and contribute together to the output of label prediction. Input image of these two networks has the same size, $1 \times 128 \times 128$. One network is responsible for learning from vibration signals and the other learns representations from current signals. The learned fault signatures from separate networks are flattened and merged to one fully connected layer with 1024 ReLUs. The output layer for predicting state label is the same as the one in architecture 1.

Both architectures are initialized randomly, and Adam optimizer is used to update the model weights with learning rate being set as 0.0001. Training dataset is used to train the

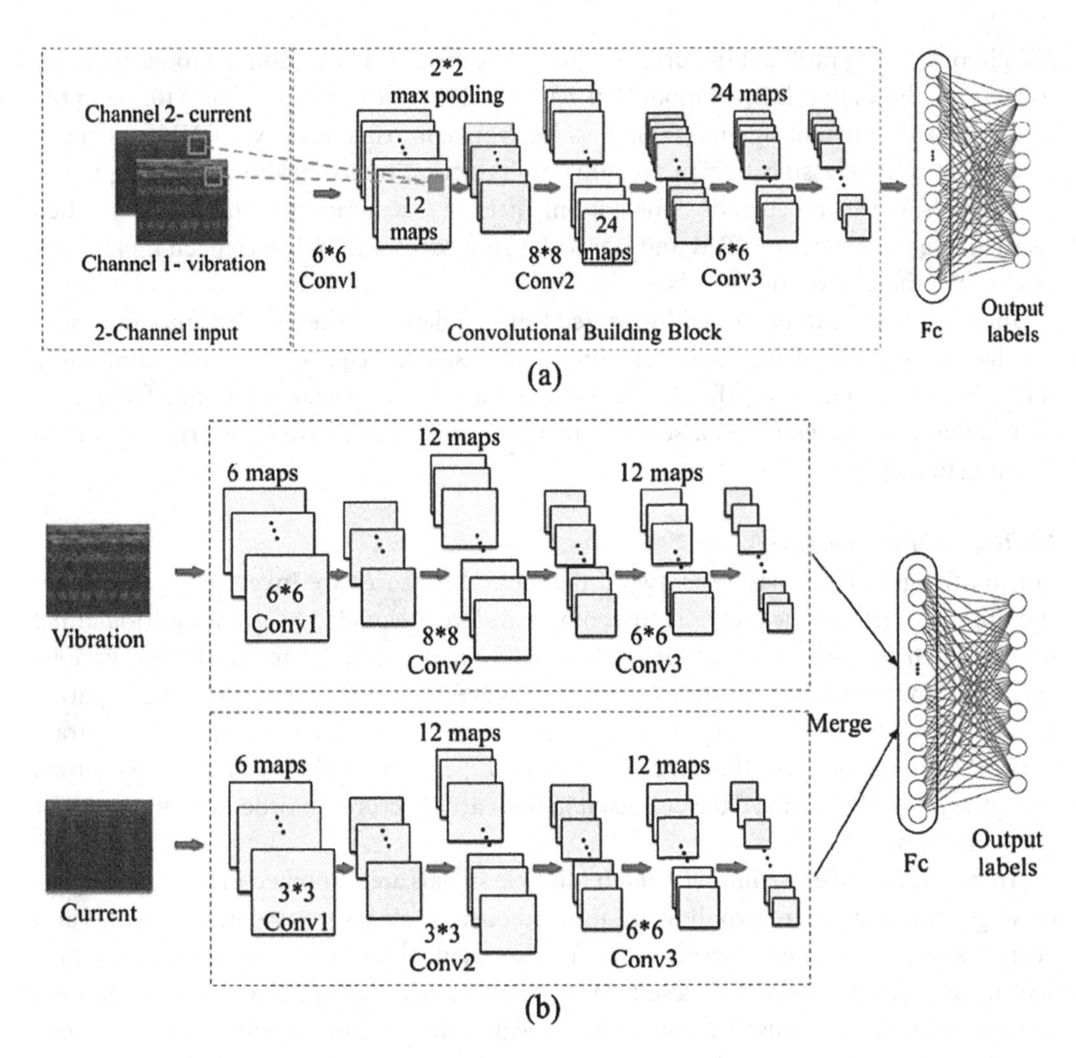

FIGURE 6.8 Multi-signal fault diagnosis with different DCNN architectures. (a) Architecture with multi-channel input, and (b) architecture with separate convolutional units and merged together in the fully connected layer.

model. Training epoch is set as 30, and the batch size is set as 32. Accuracy and training loss are calculated to evaluate the performances of each architecture, and the confusion matrix is used to show the classification accuracy for each category. In order to investigate the model stability, results from 50 trails are recorded, as shown in Table 6.3 and Figure 6.9. The confusion matrix is shown in Figure 6.10. The classification results

TABLE 6.3 Classification Results in Multi-signal Model

	Classification Accuracy Mean (Standard Deviation)		Computation Time	
	Train	Test	Train	Test
Vibration	100% (0.00)	99.22 (0.26)	135 s	0.8 s
Current	100% (0.00)	99.83 (0.16)	225 s	0.9 s

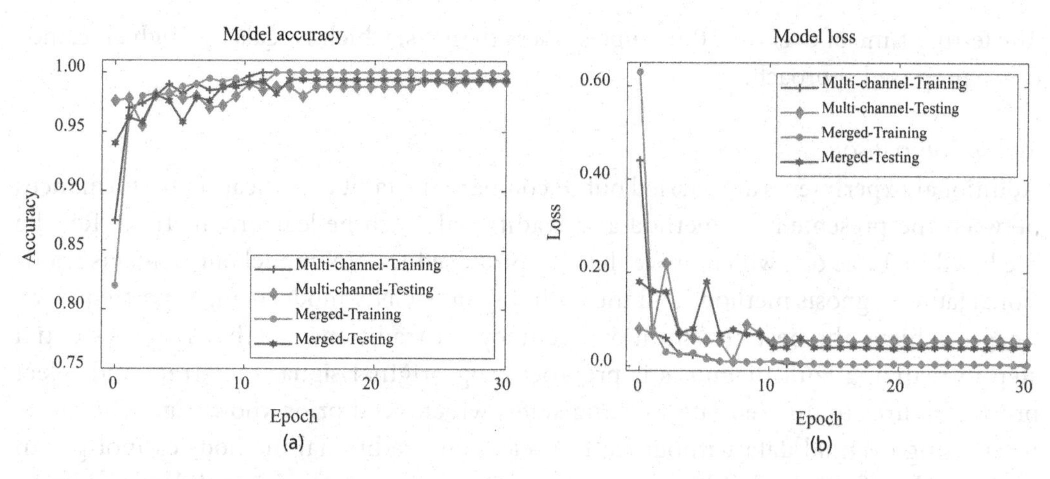

FIGURE 6.9 Multi-signal fault diagnosis model loss and classification accuracy on induction motor dataset. (a) Accuracy and (b) loss.

show that DCNN model using multiple-sensor signals has a better performance than that using the single-sensor signal. Furthermore, for two different deep architectures utilizing both vibration and current signals, the performance is slightly different. The merged model has more accurate and stable performance than the multi-channel model. This can be interpreted that when using multi-channel images, DCNN model adds features from vibration and current TFDs together at the first convolution layer which may weaken the effectiveness of unique features from each type of signal to some degree in the deeper layers. On the other hand, the merged model uses two separate convolutional units to learn from vibration TFD and current TFD separately, and the learned features are then merged together, where unique features are well conserved during forward propagation. The training time is calculated for 4800 samples, and the testing time is calculated for 1200 samples.

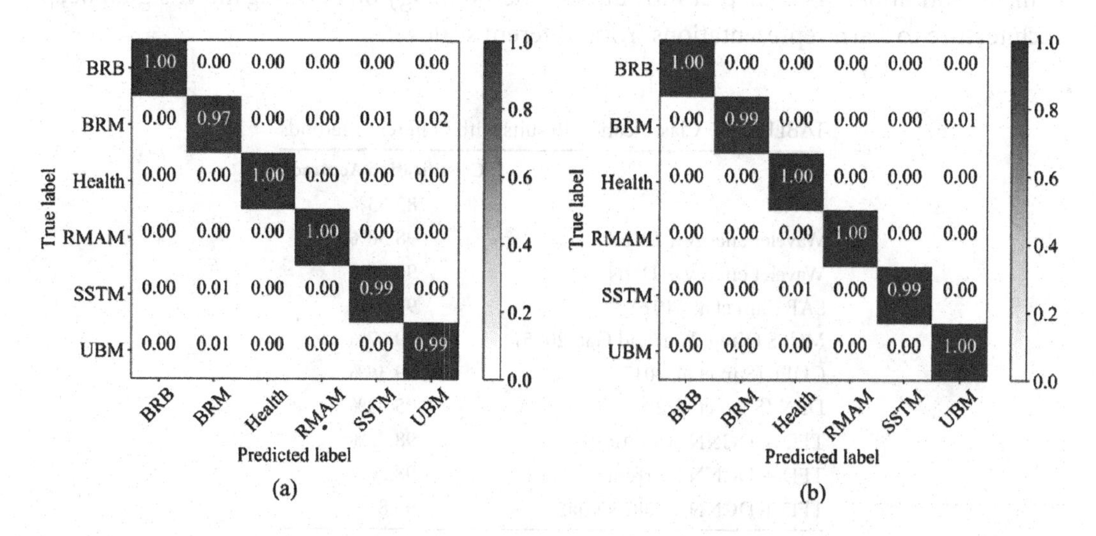

FIGURE 6.10 Confusion matrix for multi-signal model. (a) Multi-channel model and (b) merged model.

The testing time of overall 1200 samples is less than 1 s, which reveals the high efficiency of the proposed approach.

Result Comparison

Additional experiments are carried out to compare the fault classification performances between the presented DL method and traditional machine learning method. Results are listed in Table 6.4, which prove that the proposed DCNN model outperforms traditional fault diagnosis methods and the multi-signal DCNN model using time-frequency images achieve the best classification accuracy. In traditional methods, one essential step in realizing fault diagnosis is preprocessing original signal to extract and select proper features to differentiate working states which need prior knowledge. Therefore, when using original data without feature selection, traditional methods cannot get an accurate classification. For deep architecture, it can learn from original data and using the training set to adjust model parameters to automatically select features that contribute to classification accuracy where convolution operation can extract local features from the input image and different convolution kernel can learn different discriminative features. Compared with our previous works, the proposed method improves the classification accuracy for induction motor fault diagnosis. The proposed model overcomes the limitation caused by the single-sensor signal and enlarges the feature space based on multiple signals.

Different from the single-signal model, the multi-signal model is able to analyze multiple types of sensor data simultaneously and finally realize accurate fault classification based on the fused features. Each type of sensor signal contains specific information related to certain working condition. For example, the current signal has been widely used in the diagnosis of stator faults and rotor faults, while many bearing fault diagnosis methods are performed by analyzing vibration signals. Thus, combining these signals and analyzing them together is an effective alternative to realize multi-fault diagnosis even under complex conditions. This chapter introduces a methodology on building multi-signal deep architecture to learn representations from different scales.

TABLE 6.4 Classification Results with Different Methods

Model	Classification Accuracy
BPNN	82.33%
Wavelet energy + SVM	98.50%
Wavelet energy + DNN	96.83%
SAE (Sun et al. 2016)	97.61%
MCFS (Yang, Yan, and Gao 2015)	99.67%
CDFL (Sun et al. 2017)	99.36%
DBN (Shao et al. 2017)	95.33%
TFD + DCNN (vibration)	98.72%
TFD + DCNN (current)	98.26%
TFD + DCNN (multi-signal)	99.83%

Measurement Uncertainty Discussion

Due to measurement uncertainty, sensor data samples vary from each other, and using different training data samples to train deep architecture may produce various model performances. For the experimental verification, the acquired sensor data contain environmental noise and might have sensor measurement error in certain samples. In order to investigate the influence caused by measurement uncertainty, confidence interval estimation for the classification accuracy is performed using the bootstrap method (Deutsch and He 2017). The bootstrap method is a random sampling method in statistical analysis for uncertainty quantification (Endo, Watanabe, and Yamamoto 2015). For the model performance evaluation, the main procedure followed is explained below.

Initially, a resample of the original training data is generated by the bootstrap procedure. The dataset is sampled with replacement, which means each sample can be selected more than once. The size of the resample data is the same as that of the original dataset, and the samples that are not selected in the resampling are called out-of-bag samples.

Then, the proposed model is trained using the resample data and is evaluated on the out-of-bag samples. For each bootstrap iteration, a performance statistic can be obtained.

The bootstrap procedure is repeated for a number of times and a number of different resamples are generated. Repeating training and evaluation procedure, a set of performance statistics are achieved. Based on these performance statistics, a confidence interval for the designed model can be calculated.

As the number of bootstrap iterations increases, the confidence interval can be accurate. In this section, considering the number of original data and computation costs, 500 bootstrap iterations are adapted to generate classification accuracy confidence interval for the proposed multi-signal framework. The results are reported in Table 6.5 and Figure 6.11. In Figure 6.11, the color bars represent the average model classification accuracy across the bootstrap samples, and the black lines denote the confidence interval. Based on the estimation results of the confidence interval for classification accuracy, the proposed multi-signal DCNN model has relatively stable performance and different resamples have limited influence on model performance. For the merged model, there is a 95% likelihood that the range between 99.89% and 99.93% covers the model skill of fault classification. For multi-channel model, there is a 95% likelihood that the range between 99.21% and 99.29% covers its skill. Compared with the multi-channel model, the merged model has more accurate prediction and shows better capability in model robustness.

TABLE 6.5 Confidence Interval Estimation for the Classification Accuracy

	Confidence Interval		
	99%	**95%**	**90%**
Vibration	99.25 (±0.06)	99.25 (±0.04)	99.25 (±0.03)
Current	99.91 (±0.03)	99.91 (±0.02)	99.91 (±0.01)

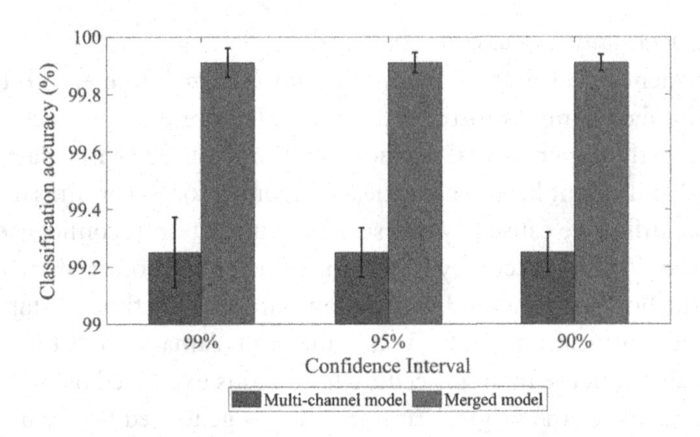

FIGURE 6.11 Confidence interval estimation for model classification accuracy.

CONCLUSION

This chapter proposes a DCNN-based multi-signal fault diagnosis framework with its application on induction motors, where TFDs of sensor signals are utilized as the input images. CNN is utilized as the building block for the multi-signal model, which shows the capability to learn discriminative features automatically from TFD images and produce accurate fault classification. Both vibration and current signals are used to verify the performance of the presented framework. Two different DCNN-based multi-signal architectures are designed and their performances are explored through experiments, where the merged model has achieved the best performance. Furthermore, the influence of measurement uncertainty is discussed based on the statistical analysis.

Future study can be conducted aiming at improving the performance of DCNN by tuning hyper-parameters and network architecture. To overcome the problem of limited training data for a deep architecture, data augmentation can be performed to enlarge the dataset. In addition, existing pre-trained models for fault diagnosis can be further investigated.

REFERENCES

Cohen, Leon. 1989. "Time-Frequency Distributions—A Review." *Proceedings of the IEEE* 77 (7): 941–981. https://doi.org/10.1109/5.30749.

Da Silva, Aderiano M, Richard J Povinelli, and Nabeel AO Demerdash. 2008. "Induction Machine Broken Bar and Stator Short-Circuit Fault Diagnostics Based on Three-Phase Stator Current Envelopes." *IEEE Transactions on Industrial Electronics* 55 (3): 1310–1318. https://doi.org/10.1109/TIE.2007.909060.

Delgado-Arredondo, Paulo Antonio, Daniel Morinigo-Sotelo, Roque Alfredo Osornio-Rios, Juan Gabriel Avina-Cervantes, Horacio Rostro-Gonzalez, and Rene de Jesus Romero-Troncoso. 2017. "Methodology for Fault Detection in Induction Motors via Sound and Vibration Signals." *Mechanical Systems and Signal Processing* 83: 568–589. https://doi.org/10.1016/j.ymssp.2016.06.032.

Deutsch, Jason, and David He. 2017. "Using Deep Learning-Based Approach to Predict Remaining Useful Life of Rotating Components." *IEEE Transactions on Systems, Man, and Cybernetics: Systems* 48 (1): 11–20. https://doi.org/10.1109/TSMC.2017.2697842.

Elbouchikhi, Elhoussin, Vincent Choqueuse, François Auger, and Mohamed El Hachemi Benbouzid. 2017. "Motor Current Signal Analysis Based on a Matched Subspace Detector." *IEEE Transactions on Instrumentation and Measurement* 66 (12): 3260–3270. https://doi.org/10.1109/TIM.2017.2749858.

Endo, Tomohiro, Tomoaki Watanabe, and Akio Yamamoto. 2015. "Confidence Interval Estimation by Bootstrap Method for Uncertainty Quantification Using Random Sampling Method." *Journal of Nuclear Science and Technology* 52 (7-8): 993–999. https://doi.org/10.1080/00223131.2015.1034216.

Glowacz, Adam, and Zygfryd Glowacz. 2017. "Diagnosis of Stator Faults of the Single-Phase Induction motor Using Acoustic Signals." *Applied Acoustics* 117: 20–27. https://doi.org/10.1016/j.apacoust.2016.10.012.

Hassan, Ola E., Motaz Amer, Ahmed K. Abdelsalam, and Barry W. Williams. 2018. "Induction motor Broken Rotor Bar Fault Detection Techniques Based on Fault Signature Analysis—A Review." *IET Electric Power Applications* 12 (7): 895–907. https://doi.org/10.1049/iet-epa.2018.0054.

Hinton, Geoffrey E., Simon Osindero, and Yee-Whye Teh. 2006. "A Fast Learning Algorithm for Deep Belief Nets." *Neural Computation* 18 (7): 1527–1554. https://doi.org/10.1162/neco.2006.18.7.1527.

Ince, Turker, Serkan Kiranyaz, Levent Eren, Murat Askar, and Moncef Gabbouj. 2016. "Real-Time motor Fault Detection by 1-D Convolutional Neural Networks." *IEEE Transactions on Industrial Electronics* 63 (11): 7067–7075. https://doi.org/10.1109/TIE.2016.2582729.

Lin, Jing, and Liangsheng Qu. 2000. "Feature Extraction Based on Morlet Wavelet and Its Application for Mechanical Fault Diagnosis." *Journal of Sound and Vibration* 234 (1): 135–148. https://doi.org/10.1006/jsvi.2000.2864.

Noureddine, Lahcène, Ahmed Hafaifa, and Abdellah Kouzou. 2017. "Rotor Fault Diagnosis of SCIG-Wind Turbine Using Hilbert Transform." 2017 9th IEEE-GCC Conference and Exhibition (GCCCE).

Rawat, Waseem, and Zenghui Wang. 2017. "Deep Convolutional Neural Networks for Image Classification: A Comprehensive Review." *Neural Computation* 29 (9): 2352–2449. https://doi.org/10.1162/neco_a_00990.

Shao, Si-Yu, Wen-Jun Sun, Ru-Qiang Yan, Peng Wang, and Robert X Gao. 2017. "A Deep Learning Approach for Fault Diagnosis of Induction Motors in Manufacturing." *Chinese Journal of Mechanical Engineering* 30 (6): 1347–1356. https://doi.org/10.1007/s10033-017-0189-y.

Sapena-Bano, Pineda Sanchez, Puche Panadero, Martinez Roman, and Dragan Matic. 2015. "Fault Diagnosis of Rotating Electrical Machines in Transient Regime using a Single Stator Current's FFT." *IEEE Transactions on Instrumentation and Measurement* 64(11): 3137–3146. https://doi.org/10.1109/TIM.2015.2444240.

Sun, Wenjun, Siyu Shao, Rui Zhao, Ruqiang Yan, Xingwu Zhang, and Xuefeng Chen. 2016. "A Sparse Auto-Encoder-Based Deep Neural Network Approach for Induction motor Faults Classification." *Measurement* 89: 171–178. https://doi.org/10.1016/j.measurement.2016.04.007.

Sun, Wenjun, Rui Zhao, Ruqiang Yan, Siyu Shao, and Xuefeng Chen. 2017. "Convolutional Discriminative Feature Learning for Induction motor Fault Diagnosis." *IEEE Transactions on Industrial Informatics* 13 (3): 1350–1359. https://doi.org/10.1109/TII.2017.2672988.

Tamilselvan, Prasanna, and Pingfeng Wang. 2013. "Failure Diagnosis Using Deep Belief Learning Based Health State Classification." *Reliability Engineering & System Safety* 115: 124–135. https://doi.org/10.1016/j.ress.2013.02.022.

Tian, Yin, Dingfei Guo, Kunting Zhang, Lihao Jia, Hong Qiao, and Haichuan Tang. 2018. "A Review of Fault Diagnosis for Traction Induction Motor." 2018 37th Chinese Control Conference (CCC).

Vincent, Pascal, Hugo Larochelle, Yoshua Bengio, and Pierre-Antoine Manzagol. 2008. "Extracting and Composing Robust Features with Denoising Autoencoders." Proceedings of the 25th International Conference on Machine Learning.

Wang, Junwei, Chuang Sun, Zhibin Zhao, and Xuefeng Chen. 2017. "Feature Ensemble Learning Using Stacked Denoising Autoencoders for Induction Motor Fault Diagnosis." 2017 Prognostics and System Health Management Conference (PHM-Harbin).

Wen, Long, Xinyu Li, Liang Gao, and Yuyan Zhang. 2017. "A New Convolutional Neural Network-Based Data-Driven Fault Diagnosis Method." *IEEE Transactions on Industrial Electronics* 65 (7): 5990–5998. https://doi.org/10.1109/TIE.2017.2774777.

Yang, Xueliang, Ruqiang Yan, and Robert X Gao. 2015. "Induction Motor Fault Diagnosis Using Multiple Class Feature Selection." 2015 IEEE International Instrumentation and Measurement Technology Conference (I2MTC) Proceedings.

CHAPTER 7

Unsupervised Deep Transfer Learning for Intelligent Fault Diagnosis

INTRODUCTION

Intelligent fault diagnosis (IFD) is becoming an essential branch among prognostic and health management (PHM) systems. While, with the increment of available data, data-driven methods with the representation learning ability are also becoming more and more important. Thus, deep learning (DL), which can extract useful features automatically from original signals, gradually becomes a hot research topic for many fields as well as PHM.

Behind the effectiveness of DL-based IFD, there exist two necessary assumptions:

1. Samples from the training dataset (source domain) should have the same distribution with that from the test dataset (target domain);
2. Plenty of labeled data are available during the training phase.

Although the labeled data might be generated by dynamic simulations or fault seeding experiments, the generated data are not strictly consistent with the test data in the real scenario. That is, DL models based on the training dataset only possess a weak generalization ability, when deployed to the test dataset from real applications. In addition, rotating machinery often operates with varying working conditions, such as loads and speeds, which also requires that trained models using the dataset from one working condition can successfully transfer to the test dataset from another working condition. In short, these factors make models trained in the source domain hard to be generalized or transferred to the target domain, directly.

Shared features existing in these two domains due to the intrinsic similarity in different application scenarios or different working conditions allow this domain shift manageable.

DOI: 10.1201/9781003474463-9

Hence, to let DL models trained in the source domain be able to be transferred well to the target domain, a new paradigm, called deep transfer learning (DTL) should be introduced into IFD. One of the effective and direct DTL is to fine-tune DL models with a few labeled data in the target domain, and then the fine-tuned model can be used to diagnose the test samples, directly. However, the newly collected data or the data under different working conditions are usually unlabeled and it is sometimes very difficult, or even impossible to label these data. Therefore, this chapter introduces the unsupervised version of DTL, called unsupervised deep transfer learning based (UDTL-based) IFD, which is to make predictions for unlabeled data on a target domain given labeled data on a source domain. It is worth mentioning that UDTL is sometimes called unsupervised domain adaptation, and this chapter does not make a strict distinction between two concepts.

In this chapter, commonly used UDTL-based settings and algorithms are discussed and a taxonomy of UDTL-based IFD is constructed. In each separate category, this chapter also gives a review about recent development of UDTL-based IFD. Some typical methods are integrated into a unified test framework, which is tested on two datasets. With this comparative study, this chapter tries to give a depth discussion (it is worth mentioning that results are just a lower bound of the accuracy) of current algorithms and attempt to discuss the core that determines the transfer performance.

The rest of this chapter is organized as follows: Section 7.2 provides background and definition of UDTL-based IFD. Basic concepts, evaluation algorithms and a brief review of UDTL-based IFD are introduced in Section 7.3. After that, in Section 7.4, datasets, evaluation results, and further discussions are provided.

RELATED BACKGROUND

This section is going to provide background and definition of UDTL-based IFD.

Definition of UDTL

To briefly describe the definition of UDTL, this section first introduces some basic symbols. It is assumed that labels in the source domain are all available, and the source domain can be defined as follows:

$$\mathcal{D}_s = \left\{\left(x_i^s, y_i^s\right)\right\}_{i=1}^{n_s}, x_i^s \in X_s, y_i^s \in Y_s, \tag{7.1}$$

where $\mathcal{D}_s$ represents the source domain, $x_i^s \in \mathbb{R}^d$ is the i-t sample, X_s is the union of all samples, y_i^s is the i-th label of the i-th sample, Y_s is the union of all different labels, and n_s means the total number of source samples. Besides, it is assumed that labels in the target domain are unavailable, and thus the target domain can be defined as follows:

$$\mathcal{D}_t = \left\{\left(x_i^t\right)\right\}_{i=1}^{n_t}, x_i^t \in X_t, \tag{7.2}$$

where $\mathcal{D}_t$ represents the target domain, $x_i^t \in \mathbb{R}^d$ is the i-th sample, X_t is the union of all samples, and n_t means the total number of target samples.

The source and target domains follow the probability distributions P and Q, respectively. We hope to build a model $\beta(\cdot)$ which can classify unlabeled samples x in the target domain:

$$\hat{y} = \beta(x), \tag{7.3}$$

where $\hat{y}$ is the prediction. Thus, UDTL is aimed to minimize the target risk $\varepsilon_t(\beta)$ using source data supervision:

$$\varepsilon_t(\beta) = \Pr_{(x,y)\sim Q}[\beta(x) \neq y]. \tag{7.4}$$

Also, the total loss of UDTL can be written as:

$$\mathcal{L} = \mathcal{L}_c + \lambda \mathcal{L}_{\mathrm{UDTL}}, \tag{7.5}$$

where $\mathcal{L}_c$ is the Softmax cross-entropy loss shown in (6), λ is the trade-off parameter, and $\mathcal{L}_{\mathrm{UDTL}}$ represents the partial loss to reduce the feature difference between source and target domains.

$$\mathcal{L}_c = -\mathbb{E}_{(x_i^s, y_i^s) \in \mathcal{D}_s} \sum_{c=0}^{C-1} \mathbf{1}_{[y_i^s = c]} \log\left[\beta\left(x_i^s\right)\right], \tag{7.6}$$

where C is the number of all possible classes, $\mathbb{E}$ denotes the mathematical expectation, and $\mathbf{1}$ is the indicator function.

Taxonomy of UDTL-Based Intelligent Fault Diagnosis

In this section, we present the taxonomy of UDTL-based IFD, as shown in Figure 7.1. We categorize UDTL-based IFD into single-domain and multi-domain UDTL according to the number of source domains from a macro perspective. In the following, we give a brief introduction of each category.

1. *Single-domain UDTL:* These can be further categorized into label-consistent (closed set) and label-inconsistent UDTL. As shown in Figure 7.2, label-consistent UDTL represents the label sets of source and target domains are consistent. Label-consistent UDTL can be classified into four categories: network-based, instanced-based, mapping-based, and adversarial-based methods from a methodological level. Additionally, we categorize label inconsistent UDTL into partial, open set, and universal tasks based on the inclusion relationship between label sets. As shown in Figure 7.2, partial UDTL means that the target label set is a subspace of the source label set; open set UDTL means that the target label set contains unknown labels; universal UDTL is a combination of the first two conditions. It is worth mentioning that three tasks can be further divided into the above four methods from a methodological level.

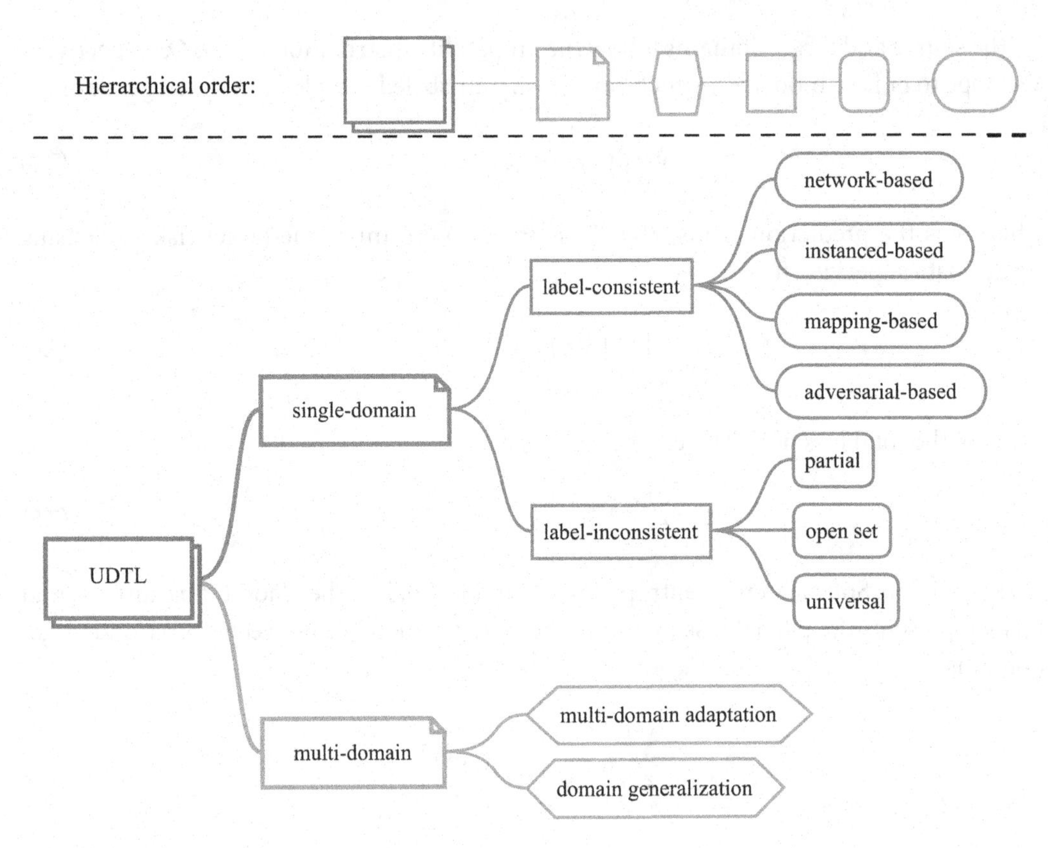

FIGURE 7.1 A taxonomy of UDTL-based methods.

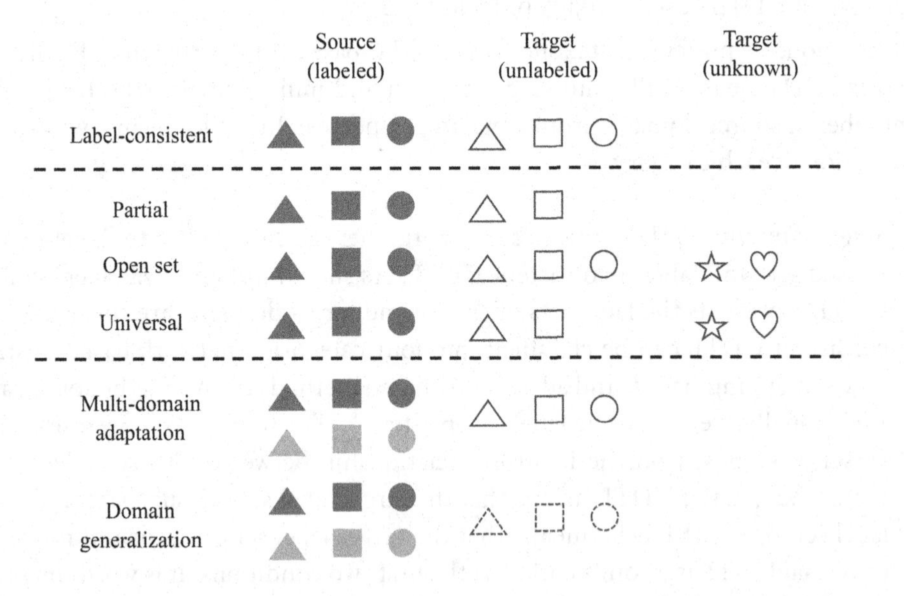

FIGURE 7.2 Visualization explanation of different transfer settings. Additionally, different colors represent different domains and dotted lines denote that this domain does not participate in training.

2. *Multi-domain UDTL:* These can be further categorized into multi-domain adaptation and domain generalization (DG) based on the usage of the target data in the training phase. Multi-domain adaptation means that the unlabeled samples from the target domain participate into the training phase, and DG is the opposite. Besides, these two conditions can also be further categorized into label-consistent and label-inconsistent UDTL.

Motivation of UDTL-Based IFD

Distributions of training and test samples are often different, due to the influence of working conditions, fault sizes, fault types, etc. Consequently, UDTL-based IFD has been introduced recently to tackle this domain shift problem since there are some shared features in the specific space. Using these shared features, applications of UDTL-based IFD can be mainly classified into four categories: Different working conditions, different types of faults, different locations, and different machines.

1. *Different working conditions*: Due to the influence of speed, load, temperature, etc., working conditions often vary during the monitoring period. Collected signals may contain domain shift, which means that the distribution of data may differ significantly under different working conditions. The aim of UDTL-based IFD is that the model trained using signals under one working condition can be transferred to signals under another different working condition.
2. *Different types of faults*: Label difference between source and target domains may exist since different types of faults would happen on the same component. Therefore, there are three cases in UDTL-based IFD. The first one is that unknown fault types appear in the target domain (open set transfer). The second one is that partial fault types of the source domain appear in the target domain (partial transfer). The third one is that the first two cases occur at the same time (universal transfer). The aim of UDTL-based IFD is that the model trained with some types of faults can be transferred to the target domain with different types of faults.
3. *Different locations*: Because sensors installed on the same machine are often responsible for monitoring different components, and sensors located near the fault component are more suitable to indicate the fault information. However, key components have different probabilities of failure rates, leading to the situation where signals from different locations have different numbers of labeled data. The aim of UDTL-based IFD is that the model trained with plenty of labeled data from one location can be transferred to the target domain with unlabeled data from other locations.
4. *Different machines*: Enough labeled fault samples of real machines are difficult to collect due to the test cost and security. Besides, enough labeled data can be generated from dynamic simulations or fault seeding experiments. However, distributions of data from dynamic simulations or fault seeding experiments are different but similar to those from real machines, due to the similar structure and measurement

situations. Thus, the aim of UDTL-based IFD is that the model can be transferred to test data gathered from real machines.

UDTL FOR INTELLIGENT FAULT DIAGNOSIS

This section is going to introduce four specific UDTL methods for IFD, including network-based, instanced-based, mapping-based, and adversarial-based methods from a methodological level.

Structure of Backbone

One of the most important parts of UDTL-based IFD is the structure of the backbone, which acts as feature extraction and has a huge impact on the test accuracy. For example, in the field of image classification, different backbones, such as VGG, ResNet, etc., have different abilities of feature extraction, leading to different classification performance.

However, for UDTL-based IFD, different studies have their own backbones, and it is difficult to determine whose backbone is better. Therefore, direct comparisons with the results listed in other published papers are unfair and unsuitable due to different representative capacities of backbones. In this section, we try to verify the performance of different UDTL-based IFD methods using the same convolutional neural network (CNN) backbone to ensure a fair comparison.

As shown in Figure 7.3, the CNN backbone consists of four one-dimension (1D) convolutional layers that come with an 1D batch normalization (BN) layer and a ReLU activation function. Besides, the second combination comes with an 1D max-pooling layer, and the fourth combination comes with an 1D adaptive max-pooling layer to realize the adaptation of the input length. The convolutional output is then flattened and passed through a fully connected (Fc) layer, a ReLU activation function, and a dropout layer. The detailed parameters are listed in Table 7.1.

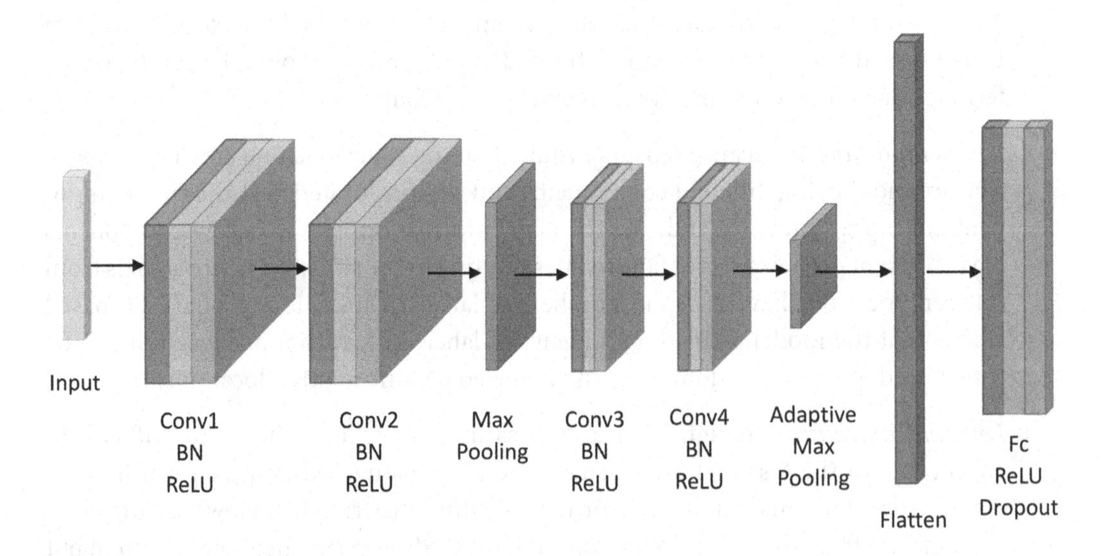

FIGURE 7.3 The structure of the backbone.

TABLE 7.1 Parameters of the Backbone

Layers	Parameters
Conv1	Out channels = 16, kernel size = 15
Conv2	Out channels = 32, kernel size = 3
Max-pooling	Kernel size = 2, stride = 2
Conv3	Out channels = 64, kernel size = 3
Conv4	Out channels = 128, kernel size = 4
Adaptive max-pooling	Output size = 4
FC	Out features = 256
Dropout	P = 0.5

Network-Based UDTL for Intelligent Fault Diagnosis

Basic Concepts

Network-based DTL means that partial network parameters pre-trained in the source domain are transferred directly to be partial network parameters of the test procedure or network parameters are fine-tuned with a few labeled data in the target domain. The most popular network-based DTL method is to fine-tune the trained model utilizing a few labeled data in the target domain. However, for UDTL-based IFD, labels in the target domain are unavailable. We use the backbone coming with a bottleneck layer, consisting of a Fc layer (out features = 256), a ReLU activation function, a dropout layer (p = 0:5), and a basic Softmax classifier to construct our basic model (we call it Basis), which is shown in Figure 7.4. The trained model is used to test samples in the target domain directly, which means that source and target domains share the same model and parameters.

Applications to IFD

Pre-trained deep neural networks using the source data were used in (Zhang et al. 2017a) via freezing their partial parameters, and then part of network parameters was transferred

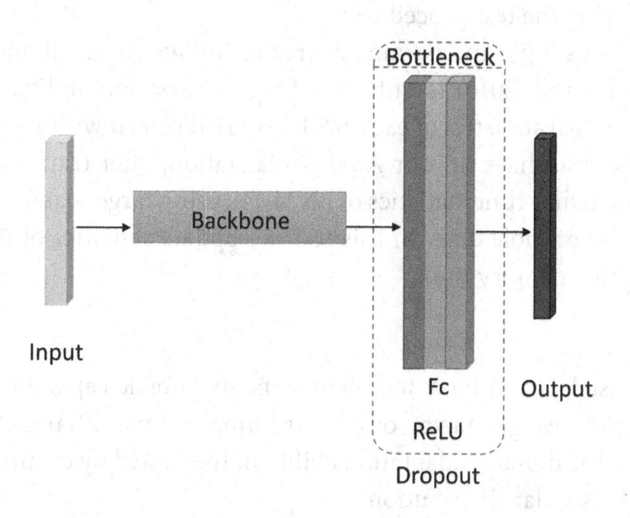

FIGURE 7.4 The structure of the basic model.

to the target network and other parameters were fine-tuned with a small amount of target data. Pre-trained deep neural networks on ImageNet were used in (Sharaf 2018) and were fine-tuned with limited target data to adapt the domain of engineering applications. Ensemble techniques and multi-channel signals were used in (He et al. 2020) to initialize the target network which was fine-tuned by a few training samples from the target domain. Two-dimensional images, such as gray images (Wang et al. 2021), time-frequency images (Wang et al. 2019), and thermal images (Shao et al. 2020), were used to pre-train the specific-designed networks, which were transferred to the target tasks via fine-tuning.

Qureshi et al. (2017) pre-trained nine deep sparse autoencoders on one wind farm, and predictions on another wind farm were taken by fine-tuning the pre-trained networks. (Zhong, Fu, and Lin 2019) trained a CNN on enough normal samples and then replaced Fc layers with support vector machine as the target model. (Han et al. 2019) discussed and compared three fine-tuning strategies: only fine-tuning the classifier, finetuning the feature descriptor, and fine-tuning both the feature descriptor and the classifier for diagnosing unseen machine conditions.

Instanced-Based UDTL for Intelligent Fault Diagnosis

Basic Concepts

Instanced-based UDTL refers to reweight instances in the source domain to assist the classifier to predict labels or use statistics of instances to help align the target domain, such as TrAdaBoost (Dai et al. 2007) and adaptive batch normalization (AdaBN) (Li et al. 2016). In this section, we use AdaBN to represent one of instanced-based UDTL methods, which does not require labels from the target domain.

BN, which can be used to avoid the issue of the internal covariate shifting, is one of the most important techniques. BN can promote much faster training speed since it makes the input distribution more stable. It is worth mentioning that BN layers are only updated in the training procedure and the global statistics of training samples are used to normalize test samples during the test procedure.

AdaBN, which is a simple and parameter-free technique for the domain shift problem, was proposed in (Li et al. 2016) to enhance the generalization ability. The main idea of AdaBN is that the global statistics of each BN layer are replaced with statistics in the target domain during the test phase. In our AdaBN realization, after training, we provide two updating strategies to fine-tune statistics of BN layers using target data, including updating via each batch and the whole data. In this test, we update statistics of BN layers via each batch considering the memory limit.

Applications to IFD

Xiao et al. (2019) used TrAdaBoost to enhance the diagnostic capability of the fault classifier by adjusting the weight factor of each training sample. Zhang et al. (2017b) used AdaBN to improve the domain adaptation ability of the model by ensuring that each layer receives data from a similar distribution.

Mapping-Based UDTL for Intelligent Fault Diagnosis

Basic Concepts

Mapping-based UDTL refers to map instances from both source and target domains to the feature space via a feature extractor. There are many methods belonging to mapping-based UDTL, such as Euclidean distance, Minkowski distance, Kullback–Leibler, correlation alignment (CORAL) (Sun and Saenko 2016), maximum mean discrepancy (MMD) (Borgwardt et al. 2006), multi-kernel MMD (MK-MMD) (Long et al. 2015), joint distribution adaptation (JDA) (Long et al. 2013), balanced distribution adaptation (BDA) (Wang et al. 2017), and joint maximum mean discrepancy (JMMD) (Long et al. 2017). In this section, we only use MK-MMD as an example to represent mapping-based methods and test their performance.

MK-MMD To introduce the definition of MK-MMD, we briefly explain the concept of MMD. MMD was first proposed by Borgwardt et al. (2006) and was used in transfer learning by many other scholars (Pan et al. 2010). MMD defined in reproducing kernel Hilbert space (RKHS) is a squared distance between the kernel embedding of marginal distributions $P(X_s)$ and $Q(X_t)$. RKHS is a Hilbert space of functions in which point evaluation is a continuous linear functional, and some examples can be found in (Żynda 2020). The formula of MMD can be written as follows:

$$\mathcal{L}_{MMD}(P,Q)=\left\|\mathbb{E}_P(\phi(x^s))-\mathbb{E}_Q(\phi(x^t))\right\|^2_{\mathcal{H}_k}, \tag{7.7}$$

where $\mathcal{H}_k$ is RKHS using the kernel k (in general, Gaussian kernel is used as the kernel), and $\phi(\cdot)$ is the mapping to RKHS.

Parameter selection of each kernel is crucial to the final performance. To tackle this problem, MK-MMD, which could maximize the two-sample test power and minimize the Type II error jointly, was proposed by (Gretton et al. 2012). For MK-MMD, scholars often use the convex combination of m kernels $\{k_u\}$ to provide effective estimations of the mapping.

$$K \triangleq \left\{k=\sum_{u=1}^{m}\alpha_u k_u : \sum_{u=1}^{m}\alpha_u=1, \alpha\geq 0, \forall u\right\}, \tag{7.8}$$

where $\{\alpha_u\}$ are weighted parameters of different kernels (in this chapter, all $\alpha_u=\frac{1}{m}$).

Inspired by deep adaptation networks (DAN) proposed in (Long et al. 2015), we design an UDTL-based IFD model by adding MK-MMD into the loss function to realize the feature alignment shown in Figure 7.5. In addition, the final loss function is defined as follows:

$$\mathcal{L}=\mathcal{L}_c+\lambda_{\text{MK-MMD}}\mathcal{L}_{\text{MK-MMD}}(\mathcal{D}_s,\mathcal{D}_t), \tag{7.9}$$

where $\lambda_{\text{MK-MMD}}$ is a trade-off parameter and $\mathcal{L}_{\text{MK-MMD}}$ means the multi-kernel version of MMD. Besides, we simply use the Gaussian kernel and the number of kernels is equal to

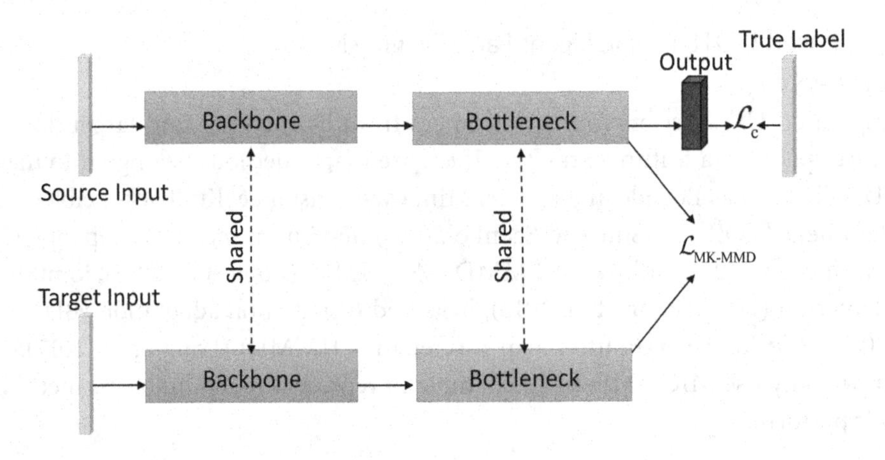

FIGURE 7.5 The UDTL-based IFD model based on MK-MMD.

five. The bandwidth of each kernel is set to be median pairwise distances on training data according to the median heuristic (Gretton et al. 2012).

Applications to IFD

BDA was used by Wang and Wu (2018) to adaptively balance the importance of the marginal and conditional distribution discrepancy between feature domains learned by deep neural networks for IFD. The CORAL loss (Wang, He, and Li 2019) and maximum variance discrepancy (MVD) (Zhang et al. 2020) were used to reduce the distribution discrepancy between different domains. (Qian, Li, and Wang 2018) considered the higher-order moments and proposed an HKL divergence to adjust domain distributions for rotating machine fault diagnosis. The distance designed to measure source and target tensor representations was proposed in (Hu, Wang, and Gu 2020) to align tensor representations into the invariant tensor subspace for bearing fault diagnosis.

Another metric distance, called MMD, was widely used in the field of intelligent diagnosis (Lu et al. 2016). (Tong et al. 2018) reduced marginal and conditional distributions simultaneously across domains based on MMD in the feature space by refining pseudo test labels for bearing fault diagnosis. Wang et al. (2020) proposed a conditional MMD based on estimated pseudo labels to shorten the conditional distribution distance for bearing fault diagnosis. The marginal and conditional distributions were aligned simultaneously in multiple layers via minimizing MMD (Li et al. 2020a).

MK-MMD was used in (Li et al. 2019) to better transfer the distribution of learned features in the source domain to that in the target domain for IFD. Han et al. (2020) used JDA to align both conditional and marginal distributions simultaneously to construct a more effective and robust feature representation for substantial distribution difference. Wang et al. (2021) further used the grey wolf optimization algorithm to learn the parameters of JDA. Based on JMMD, Cao, Chen, and Zeng (2020) proposed a soft JMMD to reduce both the marginal and conditional distribution discrepancy with the enhancement of auxiliary soft labels.

Adversarial-Based UDTL for Intelligent Fault Diagnosis

Basic Concepts

Adversarial-based UDTL refers to an adversarial method using a domain discriminator to reduce the feature distribution discrepancy between source and target domains produced by a feature extractor. In this section, we only use domain adversarial neural network (DANN) (Ganin et al. 2016) (conditional domain adversarial network (CDAN) (Zhang et al. 2018) is also recommended) to represent adversarial-based methods and test the corresponding accuracy.

DANN Similar to MMD and MK-MMD, DANN is defined to solve the problem $P(X_s) \neq Q(X_t)$. It aims to train a feature extractor, a domain discriminator distinguishing source and target domains, and a class predictor, simultaneously to align source and target distributions. That is, DANN trains the feature extractor to prevent the domain discriminator from distinguishing differences between two domains. Let G_f be the feature extractor whose parameters are θ_f, G_c be the class predictor whose parameters are θ_c, and G_d be the domain discriminator whose parameters are θ_d. After that, the prediction loss and the adversarial loss (the binary cross-entropy loss) can be rewritten as follows:

$$\begin{aligned}\mathcal{L}_c(\theta_f,\theta_c)= \\ -\mathbb{E}_{(x_i^s,y_i^s)\in\mathcal{D}_s}\sum_{c=0}^{C-1}\mathbf{1}_{[y_i^s=c]}\log\left[G_c\left(G_f\left(x_i^s;\theta_f\right);\theta_c\right)\right],\end{aligned} \tag{7.10}$$

$$\begin{aligned}\mathcal{L}_{\text{DANN}}\left(\theta_f,\theta_d\right)=-\mathbb{E}_{x_i^s\in\mathcal{D}_s}\log\left[G_d\left(G_f\left(x_i^s;\theta_f\right);\theta_d\right)\right] \\ -\mathbb{E}_{x_i^t\in\mathcal{D}_t}\log\left[1-G_d\left(G_f\left(x_i^t;\theta_f\right);\theta_d\right)\right].\end{aligned} \tag{7.11}$$

To sum up, the total loss of DANN can be defined as:

$$\mathcal{L}(\theta_f,\theta_c,\theta_d)=\mathcal{L}_c(\theta_f,\theta_c)-\lambda_{\text{DANN}}\mathcal{L}_{\text{DANN}}(\theta_f,\theta_d), \tag{7.12}$$

where λ_{DANN} is a trade-off parameter.

During the training procedure, we need to minimize the prediction loss to allow the class predictor to predict true labels as much as possible. In addition, we also need to maximize the adversarial loss to make the domain discriminator difficult to distinguish differences. Thus, solving the saddle point problem $(\theta_f,\theta_c,\theta_d)$ is equivalent to the following minimax optimization problem:

$$\begin{aligned}(\theta_f,\theta_c)=\arg\min_{\theta_f,\theta_c}\mathcal{L}(\theta_f,\theta_c,\theta_d) \\ (\theta_d)=\arg\max_{\theta_d}\mathcal{L}(\theta_f,\theta_c,\theta_d)\end{aligned} \tag{7.13}$$

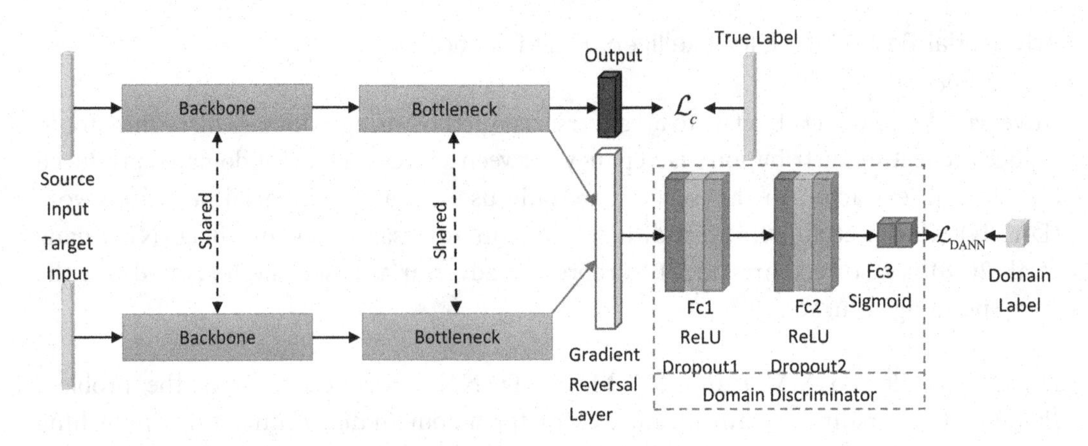

FIGURE 7.6 The UDTL-based IFD model based on DANN.

Following the statement in Ganin et al. (2016), we can simply add a special gradient reversal layer (GRL), which changes signs of the gradient from the subsequent level and is parameter-free, to solve the above optimization problem.

An UDTL-based IFD model via adding the adversarial idea into the loss function is designed to realize the feature transfer between source and target domains shown in Figure 7.6. It can be observed that a three-layer Fc binary classifier is used as the domain discriminator which is the same as (Ganin et al. 2016). The output features of these Fc layers are 1024 (Fc1), 1024 (Fc2), and 2 (Fc3), respectively. The parameter of a dropout layer is $p = 0.5$.

Applications to IFD

In Zhang et al. (2018), the feature extractor was pre-trained with the labeled source data and was used to generate target features. After that, features from source and target domains were trained to maximize the domain discriminator loss, leading to distribution alignment for IFD. Classifier discrepancy (Jiao, Zhao, and Lin 2019), which means using separate classifiers for source and target domains, was introduced in UDTL-based IFD via an adversarial training process. Meanwhile, adversarial training was also combined with other metric distances to better match the feature distributions between different domains for IFD. Li et al. (2020b) used two feature extractors and classifiers trained using MMD and domain adversarial training, respectively, and meanwhile ensemble learning was further utilized to obtain the final results. Qin et al. (2020) proposed a multiscale transfer voting mechanism (MSTVM) to improve the classical domain adaption models and the verified model was trained by MMD and domain adversarial training.

Wasserstein distance was used by Cheng et al. (2020) to guide adversarial training for aligning the discrepancy of distributions for IFD. Yu et al. (2020) combined conditional adversarial DA with a center-based discriminative loss to realize both distribution discrepancy and feature discrimination for locomotive fault diagnosis.

EXPERIMENTAL STUDIES

In this section, we test two datasets to verify the performance of different UDTL methods for comparative studies.

1. *Case Western Reserve University (CWRU) dataset:* The CWRU dataset provided by Case Western Reserve University Bearing Data Center (Case School of Engineering 2021) is one of the most famous open-source datasets in IFD and has been already used by tremendous published papers. Following other papers, this section also uses the drive end bearing fault data whose sampling frequency is equal to 12 kHz and ten bearing conditions are listed in Table 7.2. In Table 7.2, one normal bearing (NA) and three fault types including inner fault (IF), ball fault (BF) and outer fault (OF) are classified into ten categories (one health state and nine fault states) according to different fault sizes.

   Besides, as shown in Table 7.3, CWRU consists of four motor loads corresponding to four operating speeds. For the transfer learning task, this section considers these working conditions as different tasks including 0, 1, 2, and 3. For example, task 0 → 1 means that the source domain with a motor load 0 HP transfers to the target domain with a motor load 1 HP. In total, there are twelve transfer learning tasks.
2. *Paderborn University (PU) dataset:* The PU dataset acquired from Paderborn University is a bearing dataset (Lessmeier et al. 2016) which consists of artificially induced and real damages. The sampling frequency is equal to 64 kHz. Via changing the rotating speed of the drive system, the radial force onto the test bearing and the load torque on the drive train, the PU dataset consists of four operating conditions as shown in Table 7.4. Thirteen bearings with real damages caused by accelerated lifetime tests (Lessmeier et al. 2016) are used to study transfer learning tasks among different working conditions (twenty experiments were performed on each bearing code, and each experiment sustained four seconds). The categorization information is presented in Table 7.5 (the meaning of contents is explained in (Lessmeier et al. 2016)). In total, there are 12 transfer learning settings.

Experimental Description

Data Preprocessing and Splitting

Data preprocessing and splitting are two important aspects in terms of performance of UDTL-based IFD. Although UDTL-based methods often possess automatic feature learning capabilities, some data processing steps can help models achieve better performance, such as short-time Fourier transform (STFT) in speech signal classification and the data

TABLE 7.2 The Description of the Class Labels of CWRU

Class Label	**0**	**1**	**2**	**3**	**4**
Fault location	NA	IF	BF	OF	IF
Fault size (mils)	0	7	7	7	14
Class label	5	6	7	8	9
Fault location	BF	OF	IF	BF	OF
Fault size (mils)	14	14	21	21	21

TABLE 7.3 The Transfer Learning Tasks of CWRU

Task	0	1	2	3
Load (HP)	0	1	2	3
Speed (rpm)	1797	1772	1750	1730

TABLE 7.4 The Transfer Learning Tasks and Operating Parameters of PU

Task	0	1	2	3
Load torque (Nm)	0.7	0.7	0.1	0.7
Radial force (N)	1000	1000	1000	400
Speed (rpm)	1500	900	1500	1500

normalization in image classification. Besides, there often exist some pitfalls in the training phase, especially test leakage. That is, test samples are unheedingly used in the training phase.

1. Input types: There are two kinds of input types tested in this section, including the time domain input and the frequency domain input. For the former one, signals are used as the input directly and the sample length is 1024 without any overlapping. For the latter one, signals are first transformed into the frequency domain and the sample length is 512 due to the symmetry of spectral coefficients.

2. Normalization: Data normalization is the basic procedure in UDTL-based IFD, which can keep input values into a certain range. In this section, we use the Z-score normalization.

3. Data splitting: Since this section does not use the validation set to select the best model, the splitting of the validation set is ignored here. In UDTL-based IFD, data

TABLE 7.5 The Information of Bearings with Real Damages

Bearing Code	Damage	Bearing Element	Combination	Characteristic of Damage	Label
KA04	Fatigue: pitting	OR	S	Single point	0
KA15	Plastic deform: indentations	OR	S	Single point	1
KA16	Fatigue: pitting	OR	R	Single point	2
KA22	Fatigue: pitting	OR	S	Single point	3
KA30	Fatigue: pitting	OR	R	Distributed	4
KA23	Fatigue: pitting	IR(+OR)	M	Single point	5
KB24	Fatigue: pitting	IR(+OR)	M	Distributed	6
KB27	Plastic deform: indentations	OR+IR	M	Distributed	7
KI14	Fatigue: pitting	IR	M	Single point	8
KI16	Fatigue: pitting	IR	S	Single point	9
KI17	Fatigue: pitting	IR	R	Single point	10
KI18	Fatigue: pitting	IR	S	Single point	11
KI21	Fatigue: pitting	IR	S	Single point	12

OR: outer ring; IR: inner ring.
S: single damage; R: repetitive damages; M: multiple damages.

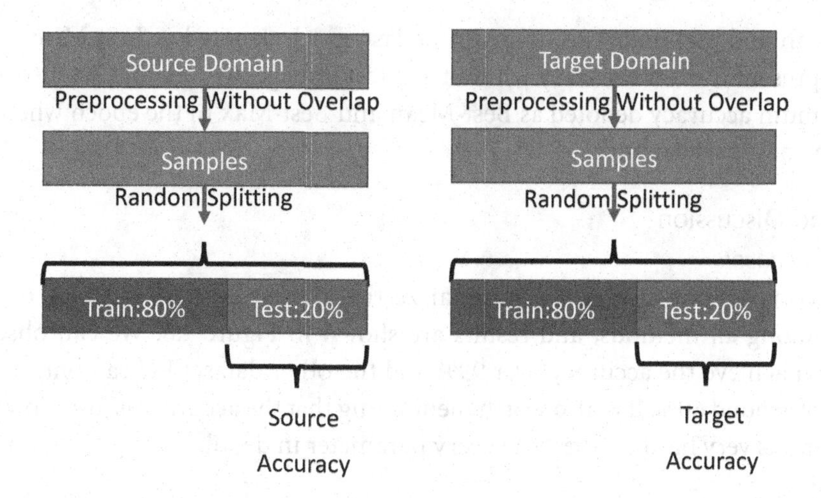

FIGURE 7.7 Data splitting for UDTL-based IFD.

in the target domain are used in the training procedure to realize the domain alignment and are also used as the test sets. In fact, data in these two situations should not overlap, otherwise there would exist test leakage. Therefore, as shown in Figure 7.7, we take 80% of total samples as the training set and 20% of total samples as the test set in source and target domains to avoid this test leakage.

Training Details

We implement all UDTL-based IFD methods in Pytorch and put them into a unified code framework. Each model is trained for 300 epochs and model training and test processes are alternated during the training procedure. We adapt minibatch Adam optimizer and the batch size is equal to 64. The "step" strategy in Pytorch is used as the learning rate annealing method and the initial learning rate is 0.001 with a decay (multiplied by 0.1) in the epoch 150 and 250, respectively. We use a progressive training method increasing the trade-off parameter from 0 to 1 via multiplying by $\frac{1-\exp(-\zeta\kappa)}{1+\exp(-\zeta\kappa)}$, where $\zeta=10$ and κ means the training progress changing from 0 to 1 after transfer learning strategies are activated.

For MK-MMD and DANN, we train models with source samples in the former 50 epochs to get a so-called pre-trained model, and then transfer learning strategies are activated. For AdaBN, we update the statistics of BN layers via each batch for three extra epochs.

All experiments are executed under Windows 10 and Pytorch 1.3 running on a computer with an Intel Core i7-9700K, GeForce RTX 2080Ti, and 16G RAM.

Evaluation Metrics

For simplicity, we use the overall accuracy, which is the number of correctly classified samples divided by the total number of samples in test data, to verify the performance of different models. To avoid the randomness, we perform experiments five times, and mean as well as maximum values of the overall accuracy are used to evaluate the final performance because variance of five experiments is not statistically useful. In this section, we use mean and maximum accuracy in the last epoch denoted as Last-Mean and Last-Max to represent the test accuracy without any test leakage. Meanwhile, we also list mean and maximum accuracy denoted as Best-Mean and Best-Max in the epoch where models achieve the best performance.

Results and Discussion

Results of Datasets

To make comparisons clearer, we summarize the highest average accuracy of different datasets among all methods, and results are shown in Figure 7.8. We can observe that CWRU can achieve the accuracy over 99% and the other dataset PU can only achieve an accuracy of around 60%. It is also worth mentioning that the accuracy is just a lower bound due to that it is very hard to fine-tune every parameter in detail.

Results of Models

Results of different methods are shown in Figure 7.9 and Figure 7.10. For the two datasets, methods discussed in this section can improve the accuracy of Basis. For AdaBN, the improvement is much smaller than other methods.

Results of DANN is generally better than those of MK-MMD, which indicates that adversarial training is helpful for aligning the domain shift.

Results of Input Types

Accuracy comparisons of two input types are shown in Figure 7.11, and it can be concluded that the time domain input achieves better accuracy in CWRU, while the frequency domain input gets better accuracy in PU. Besides, the accuracy gap between these two input types is relatively large, and we cannot simply infer which one is better due to the influence of backbones.

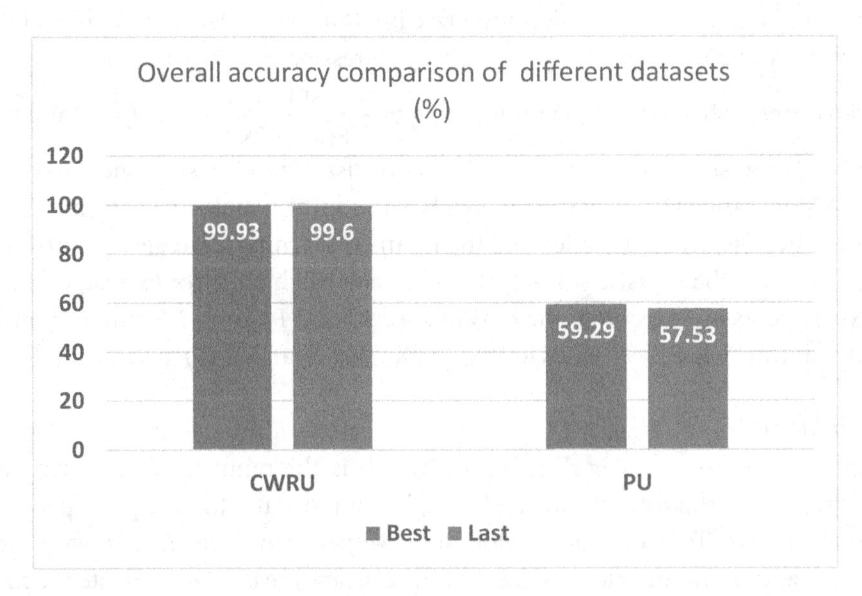

FIGURE 7.8 The highest average accuracy of the two datasets among all methods.

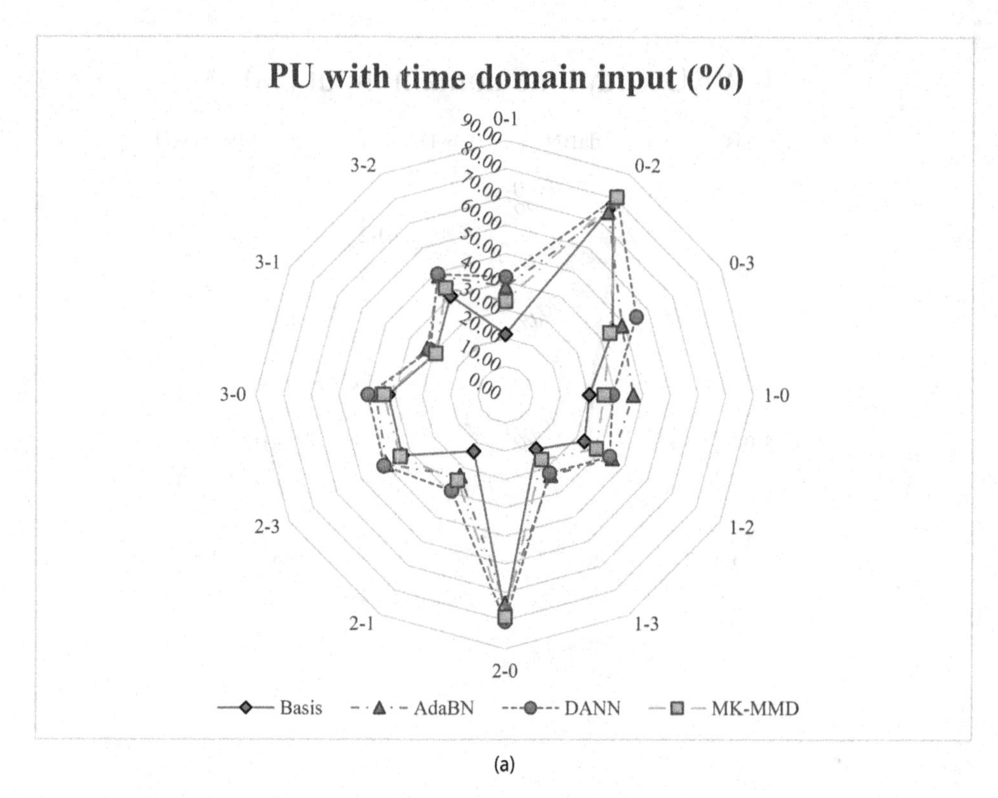

(a)

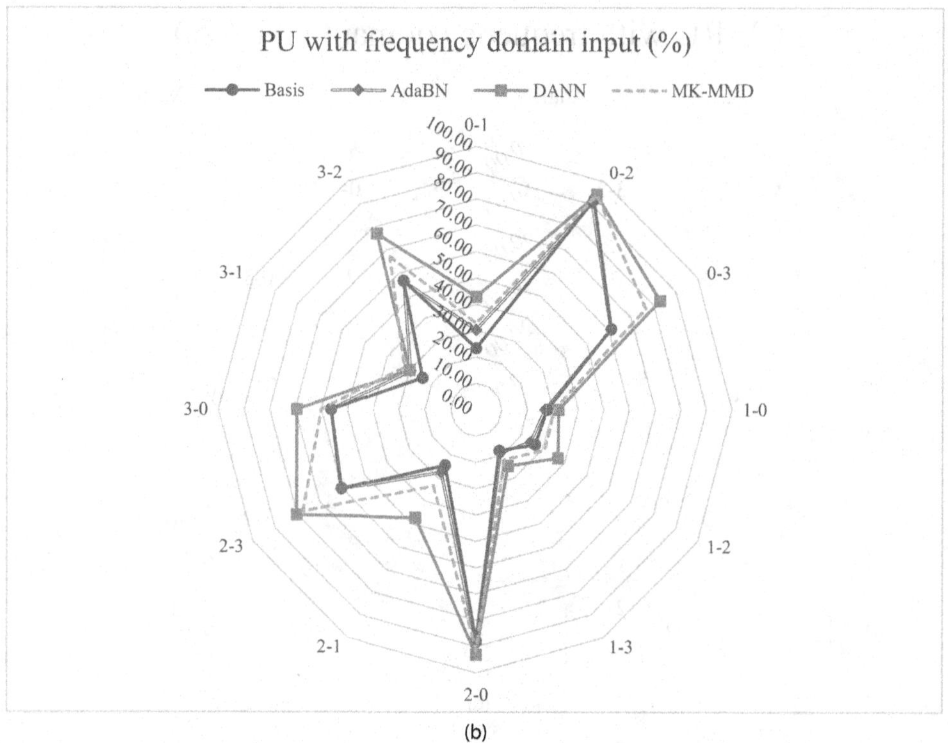

(b)

FIGURE 7.9 The accuracy comparisons of different methods in PU.

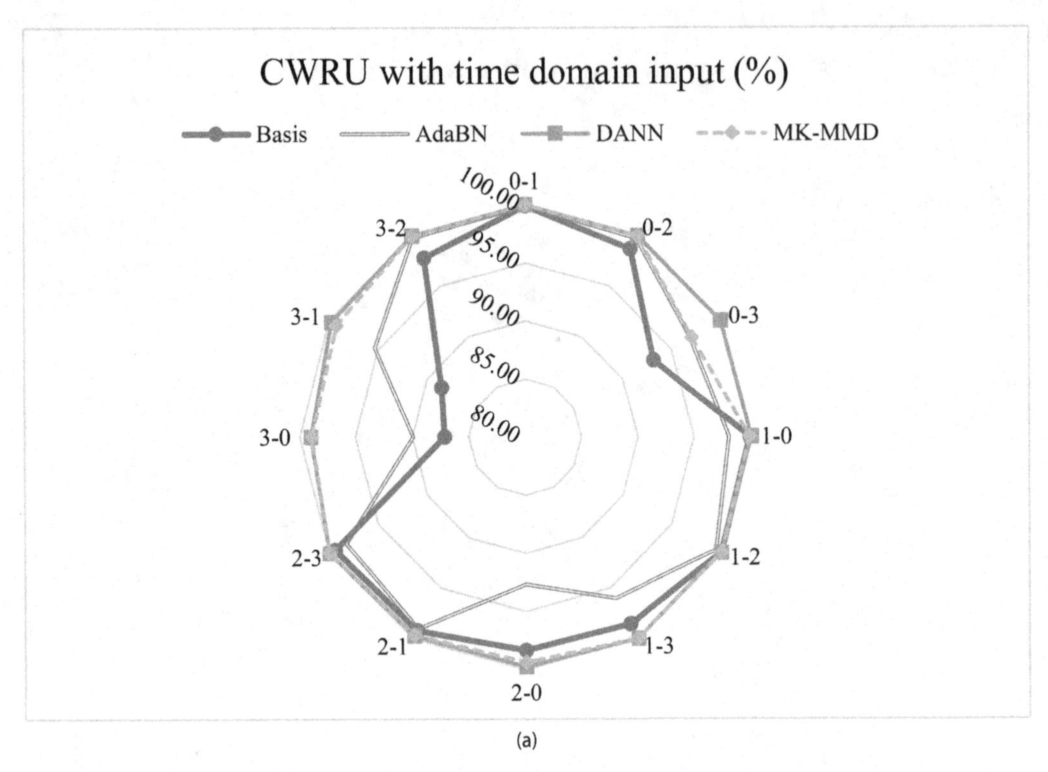

(a)

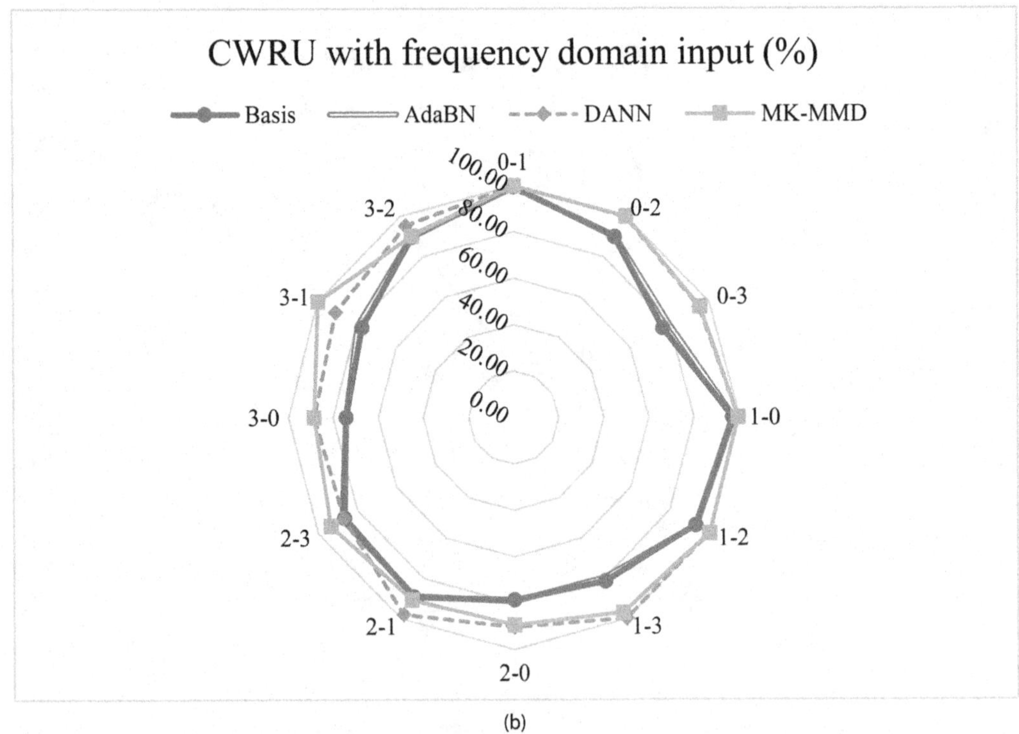

(b)

FIGURE 7.10 The accuracy comparisons of different methods in CWRU.

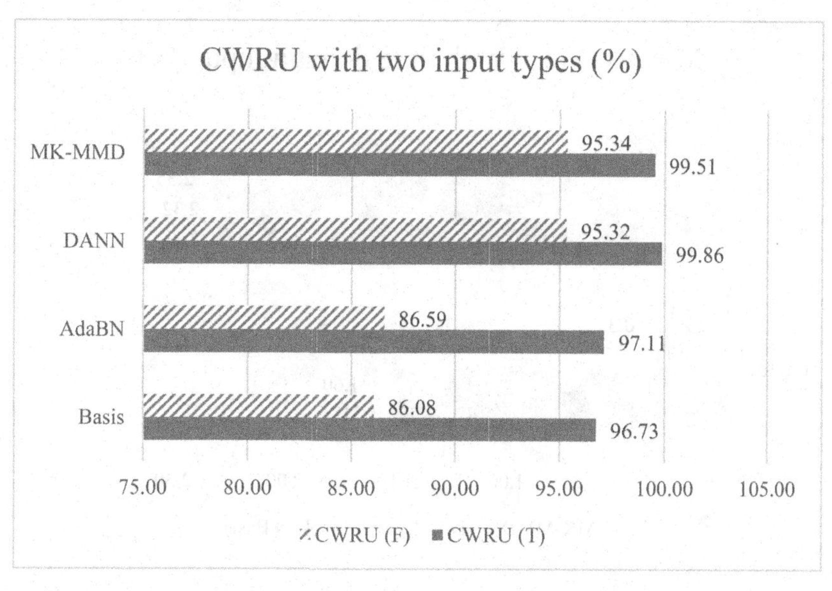

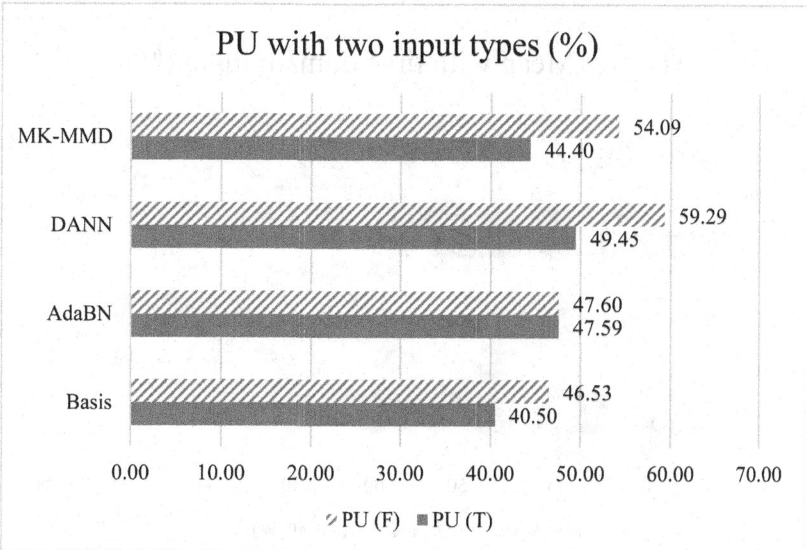

FIGURE 7.11 The accuracy comparisons of two input types with different datasets. (F) means the frequency domain input, and (T) means the time domain input.

Thus, for a new dataset, we should test results of different input types instead of just using the more advanced techniques to improve the performance of one input type, because using a different input type might improve the accuracy more efficient than using advanced techniques.

Results of Accuracy Types

We use four kinds of accuracy including Best-Mean, Best-Max, Last-Mean, and Last-Max to evaluate the performance. As shown in Figure 7.12, the fluctuation of different

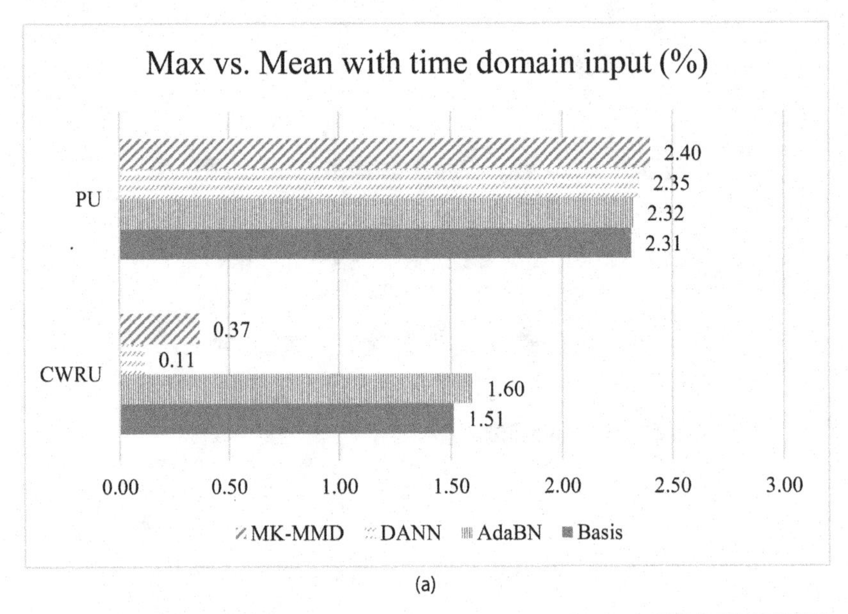

(a)

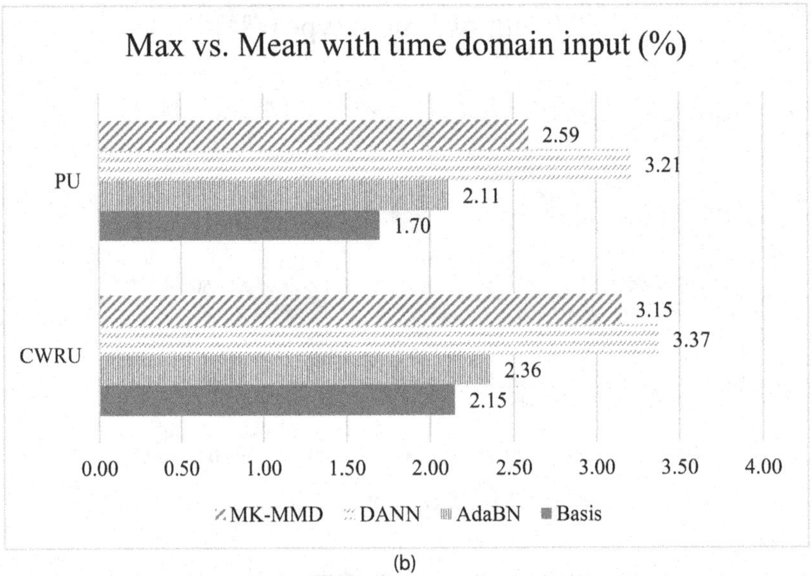

(b)

FIGURE 7.12 The difference between Max and Mean according to Best average.

experiments is sometimes large, especially for those datasets whose overall accuracy is not very high, which indicates that the used algorithms are not very stable and robust. Besides, it seems that the fluctuation of the time domain input is smaller than that of the frequency domain input, and the reason might be that the backbone used in this section is more suitable for the time domain input.

As shown in Figure 7.13, the fluctuation of different experiments is also large, which is dangerous for evaluating the true performance. Since Best uses the test set to choose the best model (it is a kind of test leakage), Last may be more suitable for representing the generalization accuracy.

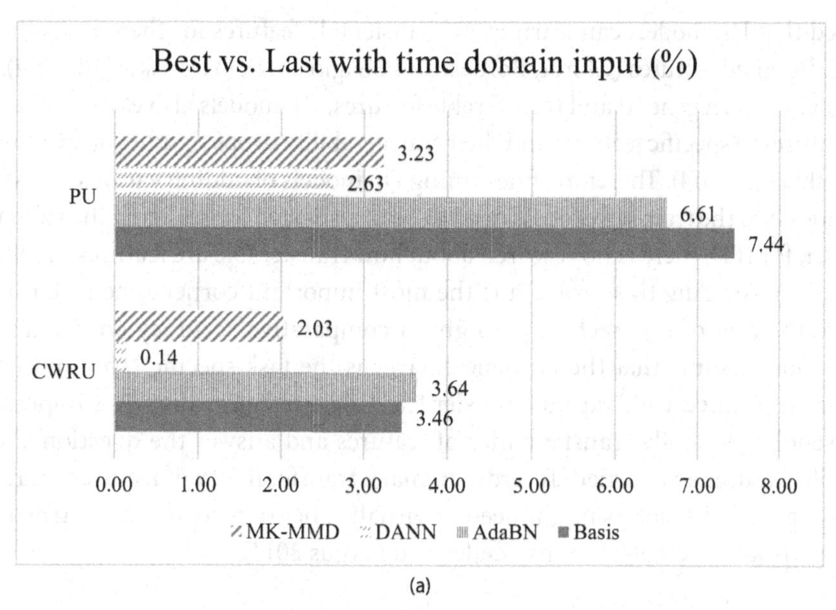

(a)

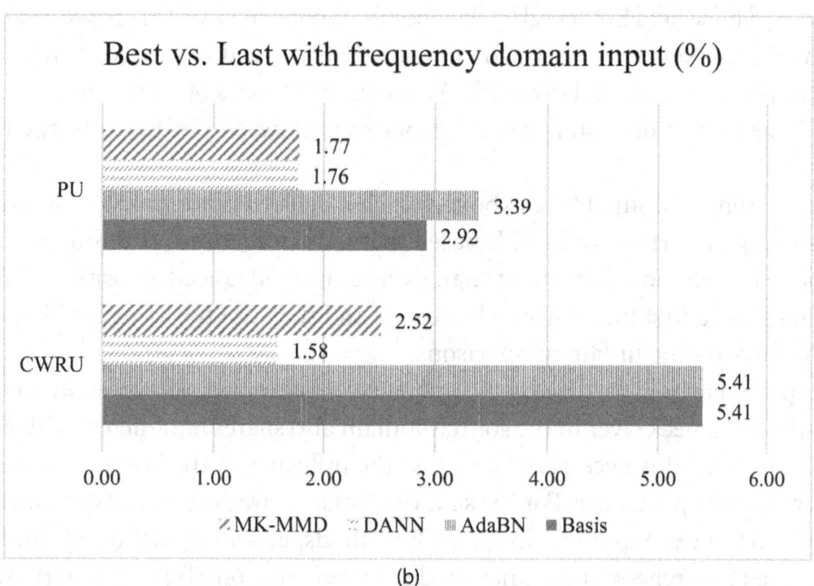

(b)

FIGURE 7.13 The difference between Best average and Last average according to Mean.

Thus, on the one hand, the stability and robustness of UDTL-based IFD need more attention instead of just improving the accuracy. On the other hand, as we analyze above, the accuracy of the last epoch (Last) is more suitable for representing the generalization ability of algorithms when the fluctuation between Best and Last is large.

Further Discussions

Transferability of Features The reason why DL models embedded transfer learning methods can achieve breakthrough performance in computer vision is that many studies have shown and proved that DL models can learn more transferable features for these tasks than traditional hand-crafted features (Glorot, Bordes, and Bengio 2011), (Yosinski et al. 2014). In spite of the ability to learn general and transferable features, DL models also exist transition from general features to specific features and their transferability drops significantly in the last layers (Yosinski et al. 2014). Therefore, fine-tuning DL models or adding various transfer learning strategies into the training process need to be investigated for realizing the valid transfer.

However, for IFD, there is no research about how transferable are features in DL models, and actually, answering this problem is the most important cornerstone in UDTL-based IFD. Since the aim of this section is to give a comparative accuracy and release a code library, we just assume that the bottleneck layer is the task-specific layer and its output features are restrained with various transfer learning strategies. Thus, it is imperative and vital for scholars to study transferability of features and answer the question about how transferable features are learned. In order to make transferability of features more reasonable, we suggest that scholars might need to visualize neurons to analyze learned features by existing visualization algorithms (Zeiler and Fergus 2014).

Influence of Backbones and Bottleneck Backbones of published UDTL-based IFD are often different, which makes results hard to compare directly, and influence of different backbones has never been studied thoroughly. Whereas, backbones of UDTL-based algorithms do have a huge impact on results from comparisons between CWRU with the frequency domain input.

Finding a strong and suitable backbone, which can learn more transferable features for IFD, is also very important for UDTL-based methods (sometimes choosing a more effective backbone is even more important than using a more advanced algorithm). We suggest that scholars should first find a strong backbone and then use the same backbone to compare results for avoiding unfair comparisons.

In the top comparison, we discuss the influence of backbones. However, in our designed structure, the bottleneck layer in the source domain also shares parameters with that in the target domain. Thus, it is necessary to discuss the influence of the bottleneck layer during the transfer learning procedure. For the sake of simplicity, we only use CWRU with two different inputs to test two representative UDTL methods, including MK-MMD and DANN.

We use Type I to represent original models in this section, Type II to represent models without the bottleneck layer, and Type III to represent models with fixed parameters of backbones (their parameters are pretrained by the source data) when starting transfer learning (only updating parameters of the bottleneck layer during the transfer learning procedure). The comparison results are shown in Figure 7.14. We can observe that for the time domain input, it is almost the same with and without the bottleneck layer. Likewise, for the frequency domain input, it is also difficult to judge which one is better. Thus, choosing a suitable network (according to datasets, transfer learning methods, input types, etc.), which can learn more transferable features, is very important for UDTL-based methods. In addition, it is clear that when parameters of backbones are fixed during the transfer learning procedure, the accuracy in the target domain decreases dramatically, which means that backbones trained using the source data cannot be transferred directly to the target domain.

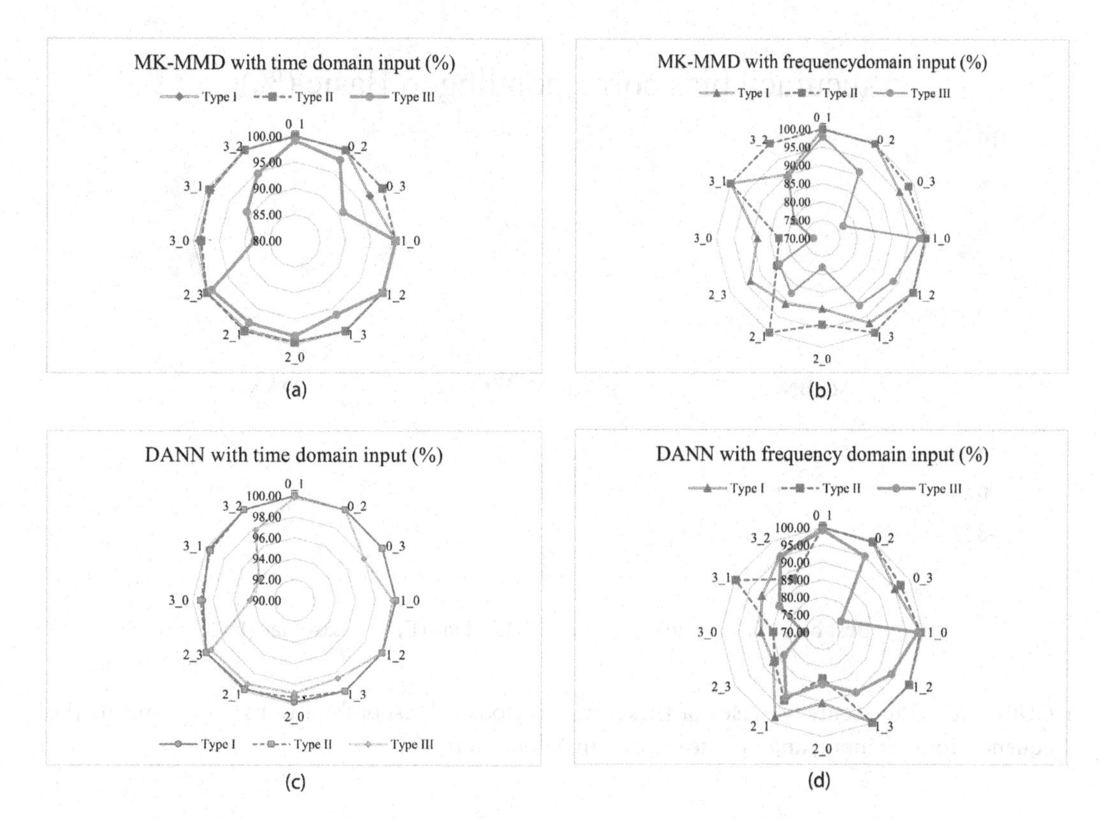

FIGURE 7.14 Comparisons of four conditions related to the bottleneck layer.

Negative Transfer There are mainly four kinds of scenarios of UDTL-based IFD, but all experiments with two datasets are about transfer between different working conditions. To state that these scenarios are not always suitable for generating the positive transfer, we use the PU dataset to design another transfer task considering the transfer between different methods of generating damages. Each task consists of three health conditions, and detailed information is listed in Table 7.6. There are two transfer learning settings in total.

The transfer results are shown in Figure 7.15. We can observe that each method has a negative transfer with the time or frequency domain inputs, and this phenomenon indicates that this constructed task may not be suitable for the transfer learning task. Actually, there are also some published papers designing transfer learning tasks which tackle transferring the gear samples to the bearing samples (it may not be a reliable transfer task) or

TABLE 7.6 The Information of Bearings with Artificial Damages

Task	Precast Method	Damage Location	Damage Extent	Bearing Code	Label
0	Electric engraver	OR	1	KA05	0
	Electric engraver	OR	2	KA03	1
	Electric engraver	IR	1	KI03	2
1	EDM and drilling	OR	1	KA01 and KA07	0
	Drilling	OR	2	KA08	1
	EDM	IR	1	KI01	2

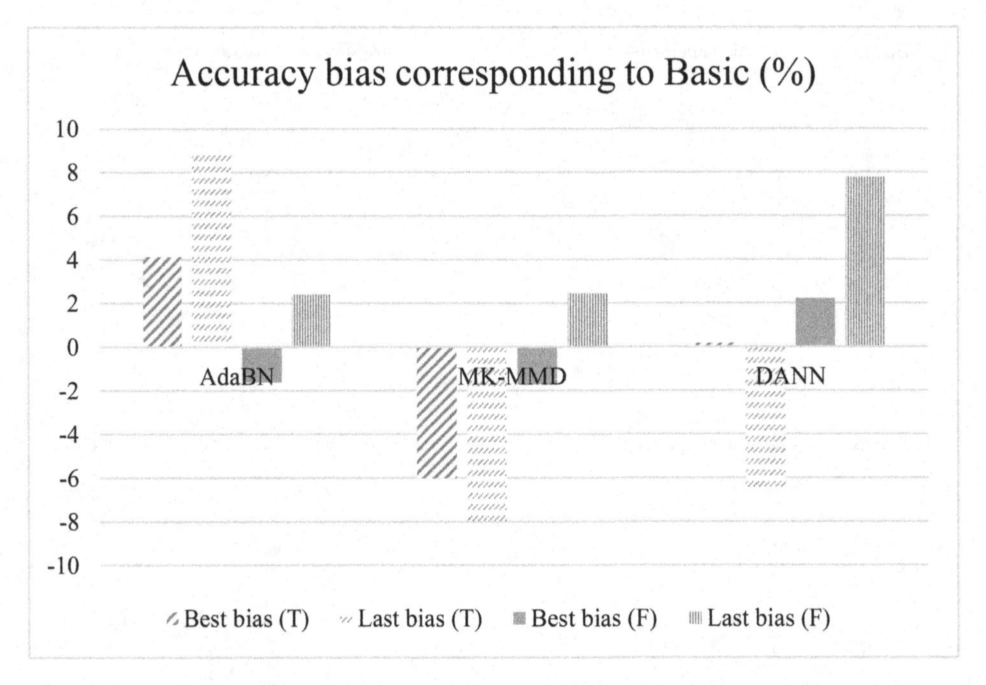

FIGURE 7.15 The accuracy biases of these five methods corresponding to basis. (F) means the frequency domain input, and (T) means the time domain input.

transferring the experimental data to the real data (if structures of two machines are different, it also may not be a reliable transfer task). Thus, it is very important to first figure out whether this task is suitable for transfer learning and whether two domains do have shared features.

Physical Priors For UDTL-based IFD, many scholars only introduce methods, which have already existed in other fields, to perform IFD tasks and pay less attention to the prior knowledge behind the data (lack of using special phenomena or rules in physical systems). Therefore, we suggest that scholars can learn from core ideas in the field of transfer learning (not just use the existing methods) and introduce prior knowledge of physical systems into the proposed method to construct more targeted and suitable diagnostic models with higher recognition rates in industrial applications.

Other Aspects Although a large amount of data in different conditions can be collected, fault data in some conditions are still scarce. Due to the fact that most machines operate in a normal condition, the class-imbalanced problem often naturally exists in real applications. Thus, imbalanced learning or few shot learning combined with transfer learning methods (Wu et al. 2020) might also be an important direction for better getting constructed algorithms off the ground.

Federated transfer learning (FTL) (Liu et al. 2020) provides a safer and more reliable approach for specific industries. At the same time, based on characteristics of transfer learning, FTL participants can own their own feature space without requiring all participants to own or use the same feature data, which makes FTL suitable for more application scenarios. FTL was initially used in IFD (Zhang and Li 2021) and more in-depth research is required.

Uncertainty quantification plays a critical role in assessing the safety of DL models during construction, optimization, and decision-making procedures. Bayesian networks (Gal and Ghahramani 2016) and ensemble learning techniques (Lakshminarayanan, Pritzel, and Blundell 2017) are two widely used uncertainty quantification methods, and their effectiveness has been verified by different kinds of applications, such as bioinformatics, self-driving car, etc. Thus, uncertainty as an auxiliary term can be used to further correct some inappropriate predictions or results during the transfer learning. For example, the prediction uncertainty is explicitly estimated during training to rectify the pseudo label learning for UDTL of semantic segmentation (Zheng and Yang 2021).

CONCLUSION

This chapter constructs a taxonomy and performs a brief review of UDTL-based IFD according to different tasks of UDTL. Two publicly available datasets are gathered to perform a comparative analysis of different UDTL-based IFD methods from several perspectives. Based on the systematically comparative study, results of different methods indicate that the assumption of joint distributions and adversarial training are two helpful techniques for promoting the accuracy. Meanwhile, different input types often behave differently on each dataset, and choosing a suitable input type might also be important to improve the accuracy. Finally, the stability and robustness of UDTL-based IFD need to be taken seriously. To sum up, it might be useful for scholars to think ahead of these results before developing new models.

REFERENCES

Borgwardt, Karsten M, Arthur Gretton, Malte J Rasch, Hans-Peter Kriegel, Bernhard Schölkopf, and Alex J Smola. 2006. "Integrating Structured Biological Data by Kernel Maximum Mean Discrepancy." *Bioinformatics* 22 (14): e49–e57. https://doi.org/10.1093/bioinformatics/btl242.

Cao, Xincheng, Binqiang Chen, and Nianyin Zeng. 2020. "A Deep Domain Adaption Model With Multi-Task Networks for Planetary Gearbox Fault Diagnosis." *Neurocomputing* 409: 173–190. https://doi.org/10.1016/j.neucom.2020.05.064.

Case School of Engineering. 2021. "Download a Data File: Case School of Engineering: Case Western Reserve University." https://engineering.case.edu/bearingdatacenter/download-data-file.

Cheng, Cheng, Beitong Zhou, Guijun Ma, Dongrui Wu, and Ye Yuan. 2020. "Wasserstein Distance Based Deep Adversarial Transfer Learning for Intelligent Fault Diagnosis With Unlabeled or Insufficient Labeled Data." *Neurocomputing* 409: 35–45. https://doi.org/10.1016/j.neucom.2020.05.040.

Dai, Wenyuan, Qiang Yang, Gui-Rong Xue, and Yong Yu. 2007. "Boosting for Transfer Learning." Proceedings of the 24th International Conference on Machine Learning.

Gal, Yarin, and Zoubin Ghahramani. 2016. "Dropout as a Bayesian Approximation: Representing Model Uncertainty in Deep Learning." International Conference on Machine Learning.

Ganin, Yaroslav, Evgeniya Ustinova, Hana Ajakan, Pascal Germain, Hugo Larochelle, François Laviolette, Mario Marchand, and Victor Lempitsky. 2016. "Domain-Adversarial Training of Neural Networks." *The Journal of Machine Learning Research* 17 (1): 2096–2030. https://jmlr.org/papers/v17/15-239.html.

Glorot, Xavier, Antoine Bordes, and Yoshua Bengio. 2011. "Domain Adaptation for Large-Scale Sentiment Classification: A Deep Learning Approach." Proceedings of the 28th International Conference on Machine Learning (ICML-11).

Gretton, Arthur, Dino Sejdinovic, Heiko Strathmann, Sivaraman Balakrishnan, Massimiliano Pontil, Kenji Fukumizu, and Bharath K Sriperumbudur. 2012. "Optimal Kernel Choice for Large-Scale Two-Sample Tests." *Advances in Neural Information Processing Systems* 25. https://dl.acm.org/doi/abs/10.5555/2999134.2999269.

Han, Te, Chao Liu, Wenguang Yang, and Dongxiang Jiang. 2019. "Learning Transferable Features in Deep Convolutional Neural Networks for Diagnosing Unseen Machine Conditions." *ISA Transactions* 93: 341–353. https://doi.org/10.1016/j.isatra.2019.03.017.

------. 2020. "Deep Transfer Network With Joint Distribution Adaptation: A New Intelligent Fault Diagnosis Framework for Industry Application." *ISA Transactions* 97: 269–281. https://doi.org/10.1016/j.isatra.2019.08.012.

He, Zhiyi, Haidong Shao, Xiang Zhong, and Xianzhu Zhao. 2020. "Ensemble Transfer CNNs Driven by Multi-Channel Signals for Fault Diagnosis of Rotating Machinery Cross Working Conditions." *Knowledge-Based Systems* 207: 106396. https://doi.org/10.1016/j.knosys.2020.106396.

Hu, Chaofan, Yanxue Wang, and Jiawei Gu. 2020. "Cross-Domain Intelligent Fault Classification of Bearings Based on Tensor-Aligned Invariant Subspace Learning and Two-Dimensional Convolutional Neural Networks." *Knowledge-Based Systems* 209: 106214. https://doi.org/10.1016/j.knosys.2020.106214.

Jiao, Jinyang, Ming Zhao, and Jing Lin. 2019. "Unsupervised Adversarial Adaptation Network for Intelligent Fault Diagnosis." *IEEE Transactions on Industrial Electronics* 67 (11): 9904–9913. https://doi.org/10.1109/TIE.2019.2956366.

Lakshminarayanan, Balaji, Alexander Pritzel, and Charles Blundell. 2017. "Simple and Scalable Predictive Uncertainty Estimation Using Deep Ensembles." *Advances in Neural Information Processing Systems* 30. https://dl.acm.org/doi/abs/10.5555/3295222.3295387.

Lessmeier, Christian, James Kuria Kimotho, Detmar Zimmer, and Walter Sextro. 2016. "Condition Monitoring of Bearing Damage in Electromechanical Drive Systems by Using Motor Current Signals of Electric Motors: A Benchmark Data Set for Data-Driven Classification." PHM Society European Conference.

Li, Qikang, Baoping Tang, Lei Deng, Yanling Wu, and Yi Wang. 2020a. "Deep Balanced Domain Adaptation Neural Networks for Fault Diagnosis of Planetary Gearboxes With Limited Labeled Data." *Measurement* 156: 107570. https://doi.org/10.1016/j.measurement.2020.107570.

Li, Xiang, Wei Zhang, Qian Ding, and Jian-Qiao Sun. 2019. "Multi-Layer Domain Adaptation Method for Rolling Bearing Fault Diagnosis." *Signal Processing* 157: 180–197. https://doi.org/10.1016/j.sigpro.2018.12.005.

Li, Yanghao, Naiyan Wang, Jianping Shi, Jiaying Liu, and Xiaodi Hou. 2016. "Revisiting batch normalization for practical domain adaptation." *arXiv preprint arXiv:1603.04779*. https://doi.org/10.48550/arXiv.1603.04779.

Li, Yibin, Yan Song, Lei Jia, Shengyao Gao, Qiqiang Li, and Meikang Qiu. 2020b. "Intelligent Fault Diagnosis by Fusing Domain Adversarial Training and Maximum Mean Discrepancy via Ensemble Learning." *IEEE Transactions on Industrial Informatics* 17 (4): 2833–2841. https://doi.org/10.1109/TII.2020.3008010.

Liu, Yang, Yan Kang, Chaoping Xing, Tianjian Chen, and Qiang Yang. 2020. "A Secure Federated Transfer Learning Framework." *IEEE Intelligent Systems* 35 (4): 70–82. https://doi.org/10.1109/MIS.2020.2988525.

Long, Mingsheng, Yue Cao, Jianmin Wang, and Michael Jordan. 2015. "Learning Transferable Features with Deep Adaptation Networks." International Conference on Machine Learning.

Long, Mingsheng, Jianmin Wang, Guiguang Ding, Jiaguang Sun, and Philip S Yu. 2013. "Transfer feature learning with joint distribution adaptation." Proceedings of the IEEE international conference on computer vision.

Long, Mingsheng, Han Zhu, Jianmin Wang, and Michael I Jordan. 2017. "Deep Transfer Learning with Joint Adaptation Networks." International Conference on Machine Learning.

Lu, Weining, Bin Liang, Yu Cheng, Deshan Meng, Jun Yang, and Tao Zhang. 2016. "Deep Model Based Domain Adaptation for Fault Diagnosis." *IEEE Transactions on Industrial Electronics* 64 (3): 2296–2305. https://doi.org/10.1109/TIE.2016.2627020.

Pan, Sinno Jialin, Ivor W Tsang, James T Kwok, and Qiang Yang. 2010. "Domain Adaptation via Transfer Component Analysis." *IEEE Transactions on Neural Networks* 22 (2): 199–210. https://doi.org/10.1109/TNN.2010.2091281.

Qian, Weiwei, Shunming Li, and Jinrui Wang. 2018. "A New Transfer Learning Method and Its Application on Rotating Machine Fault Diagnosis Under Variant Working Conditions." *IEEE Access* 6: 69907–69917. https://doi.org/10.1109/ACCESS.2018.2880770.

Qin, Yi, Xin Wang, Quan Qian, Huayan Pu, and Jun Luo. 2020. "Multiscale Transfer Voting Mechanism: A New Strategy for Domain Adaption." *IEEE Transactions on Industrial Informatics* 17 (10): 7103–7113. https://doi.org/10.1109/TII.2020.3045392.

Qureshi, Aqsa Saeed, Asifullah Khan, Aneela Zameer, and Anila Usman. 2017. "Wind Power Prediction Using Deep Neural Network Based Meta Regression and Transfer Learning." *Applied Soft Computing* 58: 742–755. https://doi.org/10.1016/j.asoc.2017.05.031.

Shao, Haidong, Min Xia, Guangjie Han, Yu Zhang, and Jiafu Wan. 2020. "Intelligent Fault Diagnosis of Rotor-Bearing System Under Varying Working Conditions With Modified Transfer Convolutional Neural Network and Thermal Images." *IEEE Transactions on Industrial Informatics* 17 (5): 3488–3496. https://doi.org/10.1109/TII.2020.3005965.

Sharaf, Sayed Ali. 2018. "Beam Pump Dynamometer Card Prediction Using Artificial Neural Networks." *KnE Engineering* 198: -212-198–212. https://api.semanticscholar.org/CorpusID: 54611662.

Sun, Baochen, and Kate Saenko. 2016. "Deep coral: Correlation alignment for deep domain adaptation." Computer Vision–ECCV 2016 Workshops: Amsterdam, The Netherlands, October 8–10 and 15–16, 2016, Proceedings, Part III 14.

Tong, Zhe, Wei Li, Bo Zhang, and Meng Zhang. 2018. "Bearing Fault Diagnosis Based on Domain Adaptation Using Transferable Features Under Different Working Conditions." *Shock and Vibration* 2018. https://doi.org/10.1155/2018/6714520.

Wang, Jianyu, Zhenling Mo, Heng Zhang, and Qiang Miao. 2019. "A Deep Learning Method for Bearing Fault Diagnosis Based on Time-Frequency Image." *IEEE Access* 7: 42373–42383. https://doi.org/10.1109/ACCESS.2019.2907131.

Wang, Jindong, Yiqiang Chen, Shuji Hao, Wenjie Feng, and Zhiqi Shen. 2017. "Balanced Distribution Adaptation for Transfer Learning." 2017 IEEE International Conference on Data Mining (ICDM).

Wang, Kaijie, and Bin Wu. 2018. "Power Equipment Fault Diagnosis Model Based on Deep Transfer Learning With Balanced Distribution Adaptation." International Conference on Advanced Data Mining and Applications.

Wang, Xiaoxia, Haibo He, and Lusi Li. 2019. "A Hierarchical Deep Domain Adaptation Approach for Fault Diagnosis of Power Plant Thermal System." *IEEE Transactions on Industrial Informatics* 15 (9): 5139–5148. https://doi.org/10.1109/TII.2019.2899118.

Wang, Xu, Changqing Shen, Min Xia, Dong Wang, Jun Zhu, and Zhongkui Zhu. 2020. "Multi-Scale Deep Intra-Class Transfer Learning for Bearing Fault Diagnosis." *Reliability Engineering & System Safety* 202: 107050. https://doi.org/10.1016/j.ress.2020.107050.

Wang, Zheng, Qingxiu Liu, Hansi Chen, and Xuening Chu. 2021. "A Deformable CNN-DLSTM Based Transfer Learning Method for Fault Diagnosis of Rolling Bearing Under Multiple Working Conditions." *International Journal of Production Research* 59 (16): 4811–4825. https://doi.org/10.1080/00207543.2020.1808261.

Wu, Jingyao, Zhibin Zhao, Chuang Sun, Ruqiang Yan, and Xuefeng Chen. 2020. "Few-Shot Transfer Learning for Intelligent Fault Diagnosis of Machine." *Measurement* 166: 108202. https://doi.org/10.1016/j.measurement.2020.108202.

Xiao, Dengyu, Yixiang Huang, Chengjin Qin, Zhiyu Liu, Yanming Li, and Chengliang Liu. 2019. "Transfer Learning With Convolutional Neural Networks for Small Sample Size Problem in Machinery Fault Diagnosis." *Proceedings of the Institution of Mechanical Engineers, Part C: Journal of Mechanical Engineering Science* 233 (14): 5131–5143. https://doi.org/10.1177/0954406219840381.

Yosinski, Jason, Jeff Clune, Yoshua Bengio, and Hod Lipson. 2014. "How Transferable Are Features in Deep Neural Networks?" *Advances in Neural Information Processing Systems* 27. https://dl.acm.org/doi/abs/10.5555/2969033.2969197.

Yu, Xiaolei, Zhibin Zhao, Xingwu Zhang, Chuang Sun, Baogui Gong, Ruqiang Yan, and Xuefeng Chen. 2020. "Conditional Adversarial Domain Adaptation With Discrimination Embedding for Locomotive Fault Diagnosis." *IEEE Transactions on Instrumentation and Measurement* 70: 1–12. https://doi.org/10.1109/TIM.2020.3031198.

Zeiler, Matthew D, and Rob Fergus. 2014. "Visualizing and Understanding Convolutional Networks." Computer Vision–ECCV 2014: 13th European Conference, Zurich, Switzerland, September 6–12, 2014, Proceedings, Part I 13.

Zhang, Bo, Wei Li, Jie Hao, Xiao-Li Li, and Meng Zhang. 2018. "Adversarial Adaptive 1-D Convolutional Neural Networks for Bearing Fault Diagnosis Under Varying Working Condition." *arXiv preprint arXiv:1805.00778*. https://doi.org/10.48550/arXiv.1805.00778.

Zhang, Ran, Hongyang Tao, Lifeng Wu, and Yong Guan. 2017a. "Transfer Learning With Neural Networks for Bearing Fault Diagnosis in Changing Working Conditions." *IEEE Access* 5: 14347–14357. https://doi.org/10.1109/ACCESS.2017.2720965.

Zhang, Wei, and Xiang Li. 2021. "Federated Transfer Learning for Intelligent Fault Diagnostics Using Deep Adversarial Networks With Data Privacy." *IEEE/ASME Transactions on Mechatronics* 27 (1): 430–439. https://doi.org/10.1109/TMECH.2021.3065522.

Zhang, Wei, Gaoliang Peng, Chuanhao Li, Yuanhang Chen, and Zhujun Zhang. 2017b. "A New Deep Learning Model for Fault Diagnosis With Good Anti-Noise and Domain Adaptation Ability on Raw Vibration Signals." *Sensors* 17 (2): 425. https://doi.org/10.3390/s17020425.

Zhang, Zhongwei, Huaihai Chen, Shunming Li, and Zenghui An. 2020. "Unsupervised Domain Adaptation via Enhanced Transfer Joint Matching for Bearing Fault Diagnosis." *Measurement* 165: 108071. https://doi.org/10.1016/j.measurement.2020.108071.

Zheng, Zhedong, and Yi Yang. 2021. "Rectifying pseudo Label Learning via Uncertainty Estimation for Domain Adaptive Semantic Segmentation." *International Journal of Computer Vision* 129 (4): 1106–1120. https://doi.org/10.1007/s11263-020-01395-y.

Zhong, Shi-sheng, Song Fu, and Lin Lin. 2019. "A Novel Gas Turbine Fault Diagnosis Method Based on Transfer Learning With CNN." *Measurement* 137: 435–453. https://doi.org/10.1016/j.measurement.2019.01.022.

Żynda, TŁ. 2020. "On Weights Which Admit Reproducing Kernel of Szegő Type." *Journal of Contemporary Mathematical Analysis (Armenian Academy of Sciences)* 55: 320–327. https://doi.org/10.3103/S1068362320050064.

CHAPTER 8

Neural Architecture Search for Intelligent Fault Diagnosis

INTRODUCTION

Fault diagnosis is an important ingredient of health monitoring system in modern industries, and it can provide discriminative information for maintenance decision making by recognizing the health status of key components in equipment. Recently, deep learning (DL) models have dominated the model building of fault diagnosis due to their ability of automatically extracting features from large-scale monitoring data (Jiang et al. 2017; Sun, Yan, and Wen 2018; Li et al. 2022). For example, Hu et al. (2020) proposed an improved network architecture to extract multiscale features for gearbox fault diagnosis. Li et al. (2020) used an adaptive channel weighted layer to rank the information importance of multisource sensors for condition monitoring of helicopter transmission system. By designing a customized network architecture manually, specific fault diagnostic task under a given domain, i.e., working condition, can be well solved.

However, multi-domain problem is common in real world scenarios, especially for the complex devices in changing environment. The data in those different domains of the corresponding key components of the complex device are inconsistent and can also be easily shifted in changing environment. For this problem, it is natural to design a matching model for data in each domain.

The automatic feature engineering of DL enables the diagnostic model to extract homologous features from the data in a given domain. But the feature quality is dependent on the architecture of DL model and the performance of one network architecture is diverse in various applicable fields. Structure determines function, which means the domain matching diagnostic model can be found by elaborately designing network architecture. However, the architecture parameters of network are complex, discrete, and disordered hyper-parameters, and required to be adjusted from multiple dimensions, such as depth, width, and identity mapping. The design space of network architecture is huge, which makes the design process a trial-and-error optimization problem.

DOI: 10.1201/9781003474463-10

Therefore, it is not feasible to design matching network architecture manually for data in each domain.

For the manual design problem, neural architecture search (NAS) is a competitive approach, which can automate the design process of network architecture for data in a given domain. NAS has attracted attention in various fields, such as computer vision (Xu et al. 2019), natural language processing (Klyuchnikov et al. 2022), and speech recognition (Baruwa et al. 2019). Scholars engaged in fault diagnosis have also gradually applied NAS methods (R.X. Wang et al. 2020). It is automation that makes it possible to achieve the matching network architecture for data in each of multiple domains. Nevertheless, conventional NAS approaches, either reinforcement learning (RL) based (Zoph and Le 2016) or evolutionary algorithm (EA) based (Real et al. 2019), are automated by resource-consuming searching, which is a discrete and heuristic process. For the sake of high efficiency, one-shot NAS methods based on hyper-network have been developed (Bender et al. 2018; Liu, Simonyan, and Yang 2018; Liang et al. 2019; Guo et al. 2019; Chu, Zhang, and Xu 2021). The hyper-network contains vast candidate network architectures, so as to avoid the cost of training the enormous amount of candidate networks from scratch. This section introduces a kind of one-shot NAS method with differentiable search strategy for fault diagnosis, which extends the differentiable architecture search (DARTS) (Liu et al. 2018) to an efficient one. The differentiable search strategy relaxes the hyper-network, and downgrades the discrete search process to a continuous optimization process, which is a resource-friendly NAS method.

Although it is efficient, DARTS presents with two drawbacks: (1) The representation of hyper-network is unsatisfactory due to the unfairness in its training and evaluation process. The coupling effect exists in the operators of hyper-network, leading to insufficient training for parametric operators and the co-adaptation problem. Besides, the sampling matrix of hyper-network is constrained by the Softmax normalization, causing Matthew effect in evaluation process. (2) The searched architecture of DARTS is generally an overconfident result, which is just one of a set with multiple high-performance architectures and not always the optimal (Zhang, Zhang, and Yang 2020; Li et al. 2020). However, it is not feasible for DARTS to get a candidate set during one searching phase.

To overcome the above two drawbacks of DARTS, this section develops a one-shot NAS method, that is, Bayesian differentiable architecture search (BDAS). The two-stage optimization is taken instead of alternating optimization for flexibility (Guo et al. 2019; Chu et al. 2021). For the unfairness problem, the hyper-network is trained based on warmup and path-dropout in the first stage, then the sampling matrix is optimized without the constraint of the softmax normalization in the second stage. For the overconfident result problem, inspired by Bayesian deep learning (Wang and Yeung 2016) and variational autoencoder (Kingma and Welling 2013), each point of the sampling matrix is extended as a distribution through variational Bayesian inference, then the candidate set can be obtained by forward sampling from the posterior distributions group. It is worth noting that, to satisfy the binarization of the sampling distributions group, the regularization of the maximum a posterior (MAP) estimation is generalized into our objective function derived based on the Kullback-Leibler (KL) divergence. The contributions of this work are listed as follows:

1. A one-shot NAS approach named BDAS is proposed to automatically design domain matching diagnostic model.
2. Intuitive factors of the unfairness of hyper-network are analyzed, and a corresponding strategy based on warmup and path-dropout is proposed for the training phase of BDAS. Then in the customized phase, variational Bayesian inference is used to estimate the uncertainty of model matching, and a scale mixture prior is used to softly constrain the matching stage.

With these contributions, experiments were performed on the aeroengine bevel gear under multiple working conditions. Different energy levels of Gaussian white noises are added into the vibration signal to shift the raw data domain, so as to simulate the multi-domain problem. Result showed that our proposed method outperforms other manually designed networks and solves the two drawbacks of DARTS.

The rest is organized as follows. Section 8.2 reviews some related works to make our motivation clear and describes the preliminary about our proposed method. In Section 8.3, the proposed BDAS method is described in detail. Experimental results and analysis are illustrated in Section 8.4. Finally, Section 8.5 summarizes the main conclusions.

RELATED BACKGROUND

One-Shot NAS

Recently, one-shot NAS methods based on the hyper-network (Liu et al. 2018; Bender et al. 2018; Liang et al. 2019; Guo et al. 2019; Chu et al. 2021) have attracted much attention due to its resource-friendly characteristics. The hyper-network provides performance proxy for the candidate networks, then the flexile search strategy, e.g., RL or EA, can be used to rank the search space. If the hyper-network can't reflect the ground truth ranking of the search space, it will propagate bias to the subsequent searching phase. In fact, a variety of training strategies have been proposed to alleviate this problem. DARTS pre-trains the hyper-network before the alternating optimization. DARTS+ (Liang et al. 2019) adopts early stopping to avoid the aggregation of skip connection caused by the unfairness of hyper-network. However, this global training causes the coupling effect, which means the parametric operators get insufficient training as the hyper-network prefers the nonparametric operators with stable inchoate performance. Besides, the Softmax normalization in the search phase of DARTS leads to the winner-takes-all evaluation, then the Matthew effect and the coupling effect will further amplify the bias under alternating optimization.

Therefore, the local training strategies for the hyper-network have been researched to replace the global training (Bender et al. 2018; Guo et al. 2019; Chu et al. 2021). SPOS (Guo et al. 2019) proposed a uniform sampling strategy to train the hyper-network, and in this method only one sampling path will be updated. FairNAS (Chu et al. 2021) proposed a fair sampling strategy inspired from the uniform sampling, which carried out multiple sampling without replacement to get a syncretic loss to update the hyper-network during once iteration. The method for the hyper-network training that most resembles our work is one-shot (Bender et al. 2018), which proposed a path-dropout strategy where several paths were dropped randomly in each update. However, this method assumes that the initial performance of all operators is equivalent. Our method warms up the parametric operators to initialize their performance to a certain level, and our strategy will further enhance the representation of the hyper-network.

It is worth noting that SPOS and FairNAS both take RL or EA as the search strategy instead of the differentiable search strategy, as DARTS still has another drawback. It is not feasible for the differentiable search strategy to get a candidate architectures set during one searching phase, while RL-based or EA-based NAS methods can easily obtain such set. Besides, DARTS tends to search a relatively good architecture rather than the optimal ones (Zhang et al. 2020). Therefore, by expanding the range of searched architectures, the better ones are expected to be obtained. Our method applies the variational Bayesian inference to obtain such candidate architectures set. To our best knowledge, it is the first work to obtain such set under the differentiable search strategy.

DARTS

Instead of searching the global network architecture, DARTS is a cell-based searching method, and its entire network is composed of a number of repetitive cell structures. The cell structure is usually represented as a directed acyclic graph (DAG) with several nodes and edges. These nodes include input/output nodes and intermediate nodes, which represent different features. The input nodes are the features copied from the previous cell. These edges include fourteen mixed edges and four connecting edges, which represent the operators between nodes. The key point of DAG is the construction of mixed edges, each of which is superimposed by eight operators in the search space multiplied by a coefficient vector called architecture weight, denoted as $\boldsymbol{\alpha}^j = \left(\alpha_1^j, \alpha_2^j, \ldots, \alpha_8^j\right)$, where j is the number of the fourteen mixed edge. Then all the fourteen vectors are spliced into a sampling matrix. So the dimension of the matrix is 14×8. Corresponding to the design of model structure, there are two DAG types, that is, normal cell and reduce cell. Since there is no essential difference between normal cell and reduce cell in design and optimization, we shall take one DAG as instance for simplicity.

The forward computation of the DAG can be regarded as the interaction between the sampling matrix and the search space:

$$f_{DAG} = \underbrace{\begin{bmatrix} \alpha_1^1 & \cdots & \alpha_8^1 \\ \vdots & \ddots & \vdots \\ \alpha_1^{14} & \cdots & \alpha_8^{14} \end{bmatrix}}_{A} \otimes \underbrace{\begin{bmatrix} OP_1(x) \\ \vdots \\ OP_8(x) \end{bmatrix}}_{Search\ space} \tag{8.1}$$

where OP_i represents the i-th operator in the defined search space, which will be described in Section IV. All the $\boldsymbol{\alpha}^j$ vectors are spliced into a sampling matrix, denoted as $\boldsymbol{A}$, and each row of this matrix represents a mixed edge. The symbol ⊗ is defined as the interaction of the matrix and the search space.

The searching process can be regarded as a bi-level optimization problem, as shown in the following equation:

$$\begin{aligned} A^* &= \arg\ \min_A L_{val}(w^*(A), A) \\ \text{s.t. } w^*(A) &= \arg\ \min_w L_{train}(w, A) \end{aligned} \tag{8.2}$$

where w^* is the convolutional kernel weights of OP under fixed $\boldsymbol{A}$, and $\boldsymbol{A}^*$ denotes the optimized sampling matrix. The search phase of DARTS is reduced to the optimization of the matrix $\boldsymbol{A}$. The final searched architecture is derived by selecting the operator with the maximum coefficient from each row of the matrix.

BDAS-BASED INTELLIGENT FAULT DIAGNOSIS

A schematic overview of our proposed BDAS method for domain matching fault diagnosis is described in Figure 8.1. It consists of four steps: data acquisition, training the hyper-network, searching the matching network set, and inference. The aeroengine bevel gear fault data in the testing bench are collected, and divided into train dataset, validation dataset, and test dataset. The hyper-network is trained based on warmup and path-dropout to obtain an intuitively fair proxy. After that, the variational inference is used to extend each point of the sampling matrix to a distribution and a scale mixture prior is used to softly constrain the matching stage. Finally, the matching network can be sampled and ordered from the posterior distributions group by analyzing the inference results.

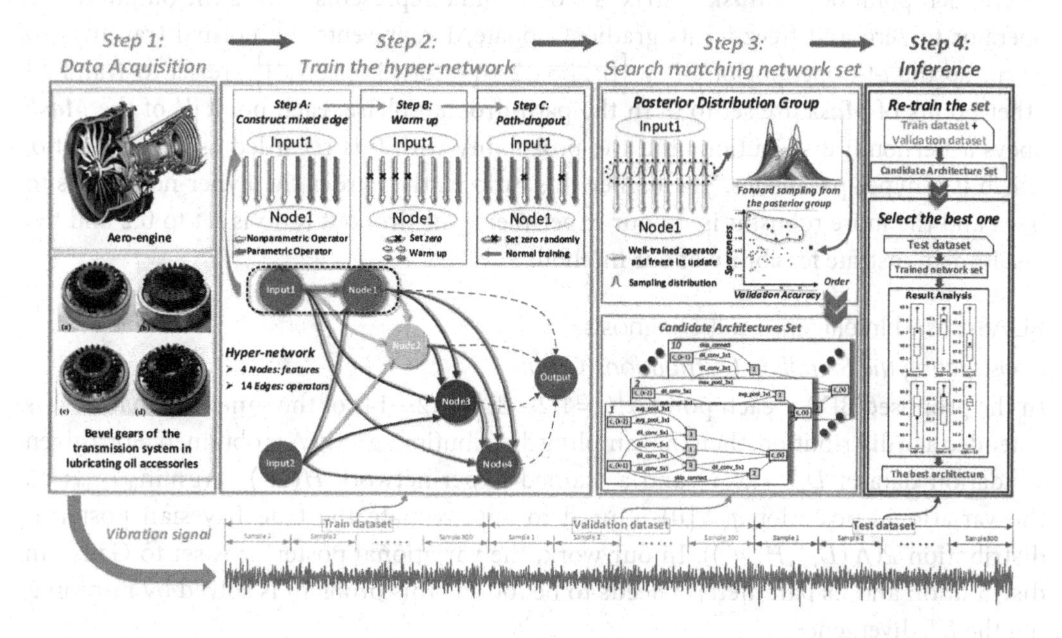

FIGURE 8.1 An overview of the proposed method for domain matching fault diagnosis.

Training Hypernetwork by Warmup and Path-Dropout

Corresponding to the search space, each mixed edge of the DAG includes four nonparametric operators $OP_{1\text{-}4}$ and four parametric operators $OP_{5\text{-}8}$. The inchoate performance of $OP_{5\text{-}8}$ is less stable than that of $OP_{1\text{-}4}$ due to the random initialization. The hyper-network is easy to fall into the trap of shortcut learning, leading to insufficient training for $OP_{5\text{-}8}$. For example, skip connection is a kind of shortcut for the hyper-network training compared with the convolutional layer. Then the performance of $OP_{5\text{-}8}$ will be seriously underestimated in the subsequent searching phase, thus resulting in the aggregation of skip connection. But the information provided by the well-trained convolutional layer is more valuable than skip connection. Therefore, we warm up each of $OP_{5\text{-}8}$ separately to enhance their competitiveness.

In addition to the difference of operators, the coupling effect exists in the training of hyper-network. The co-adaptation among operators intensifies the insufficient training degree of $OP_{5\text{-}8}$. Therefore, path-dropout is taken into the hyper-network training to alleviate this co-adaptation.

Two mask matrixes are used to realize the warmup and path-dropout, respectively, and their function can be expressed in the same formula as:

$$f_{DAG}(x)=\underbrace{\begin{bmatrix} \beta_1^1 & \cdots & \beta_8^1 \\ \vdots & \ddots & \vdots \\ \beta_1^{14} & \cdots & \beta_8^{14} \end{bmatrix}}_{\textbf{\textit{Mask}}} \otimes \underbrace{\begin{bmatrix} OP_1(x) \\ \vdots \\ OP_8(x) \end{bmatrix}}_{\textit{Search space}} \tag{8.3}$$

where each point of the ***Mask*** matrix is 0 or 1, and 0 represents setting the output of the operator to zero and freezing its gradient update, 1 represents the normal training. To warm up the i-th operator of $OP_{5\text{-}8}$, $\left\{\beta_i^j, j=1,2\cdots14, \text{ for } i \text{ in } [5,6,7,8]\right\}$ are set to 1 and all other points of ***Mask*** are set to 0. In the path-dropout phase, each point β_i^j of the ***Mask*** obeys a Bernoulli distribution, and the probability with 0 is regarded as dropout ratio, which is a hyper-parameter. The higher this ratio is, the harder the hyper-network is to train and the more robust it is. In our experiment, the dropout ratio is set to 0.2 and the results demonstrate feasibility of our method.

BDAS-Based Intelligent Fault Diagnosis

Constructing the Sampling Distributions Group

In the proposed BDAS, each point $\alpha_i^j (i=1,2\cdots8; j=1,2\cdots14)$ of the sampling matrix $\boldsymbol{A}$ is extended as a distribution, then the sampling distributions group Λ is obtained. For a given validation dataset $D_{val}=(x,y)$ and a trained hyper-network $H(w^*)=\arg\min_w L_{train}(w)$, the variational posterior $q(\Lambda|\theta)$ is used to approximate the true Bayesian posterior distribution $P(\Lambda|D_{val},H(w^*))$. In our work, the variational posterior is set to Gaussian distribution and its parameter θ needs to be found. This problem is solved by minimizing the KL divergence:

$$
\begin{aligned}
\theta^* &= \arg\min_\theta KL[q(\Lambda|\theta)||P(\Lambda|D_{val},H(w^*))] \\
&= \arg\min_\theta \int q(\Lambda|\theta)\log\frac{q(\Lambda|\theta)}{P(\Lambda|D_{val},H(w^*))}d\Lambda \\
&= \arg\min_\theta \int q(\Lambda|\theta)\log\frac{q(\Lambda|\theta)P(D_{val},H(w^*))}{P(\Lambda)P(D_{val},H(w^*)|\Lambda)}d\Lambda \\
&= \arg\min_\theta \int q(\Lambda|\theta)\log\frac{q(\Lambda|\theta)}{P(\Lambda)P(D_{val},H(w^*)|\Lambda)}d\Lambda \\
&+P(D_{val},H(w^*))\int q(\Lambda|\theta)d\Lambda \\
&= \arg\min_\theta \int q(\Lambda|\theta)\log\frac{q(\Lambda|\theta)}{P(\Lambda)P(D_{val},H(w^*)|\Lambda)}d\Lambda.
\end{aligned}
\tag{8.4}
$$

where $P(\Lambda)$ is the prior. This cost function is known as the expected lower bound. We follow the work of Blundell et al. (2015), who proposed an approximate form for the expected lower bound to allow more combinations of prior and variational posterior families. So we approximate Eq. (8.4) as:

$$
\theta^* \approx \arg\min_\theta \sum_{k}^{n}\sum_{j=1}^{14}\sum_{i=1}^{8}\log q\left((\Lambda_i^j)^k|\theta\right)-\log P\left((\Lambda_i^j)^k\right)-\log P\left(D_{val},H(w^*)|(\Lambda_i^j)^k\right) \tag{8.5}
$$

where $(\Lambda_i^j)^k$ denotes the kth Monte Carlo sample from the distribution in the posterior group, n is the Monte Carlo sampling number and n is set to 1 in the implement. The order of summation in Eq. (8.5) has been switched as they are independent and finite. In our experimental study, we found this approximate form can adequately achieve our purpose.

The reparameterization trick is used to calculate the unbiased Monte Carlo gradient for the optimization of Eq. (8.5). Each variational posterior in our work is set to a Gaussian distribution with a mean μ_i^j and a standard deviation σ_i^j. The standard deviation is parameterized as $\sigma_i^j = \log(1+\exp(\rho_i^j))$ to be positive. So the parameters $\theta_i^j = (\mu_i^j, \rho_i^j)$ need to be estimated. Then a sample of Λ can be obtained from a unit Gaussian and the transform is: $\hat{\Lambda}_i^j = \mu_i^j + \log(1+\exp(\rho_i^j)) \circ \varepsilon_i^j$ where $\hat{\Lambda}_i^j$ is a sampling instance from Λ_i^j, ε_i^j is sampled from a unit Gaussian and $\circ$ represents pointwise multiplication. The forward calculation of the DAG is:

$$
f_{DAG}(x) = \underbrace{\left\{\begin{bmatrix} \mu_1^1 & \cdots & \mu_8^1 \\ \vdots & \ddots & \vdots \\ \mu_1^{14} & \cdots & \mu_8^{14} \end{bmatrix} + \begin{bmatrix} \sigma_1^1 & \cdots & \sigma_8^1 \\ \vdots & \ddots & \vdots \\ \sigma_1^{14} & \cdots & \sigma_8^{14} \end{bmatrix} \circ \begin{bmatrix} \varepsilon_1^1 & \cdots & \varepsilon_8^1 \\ \vdots & \ddots & \vdots \\ \varepsilon_1^{14} & \cdots & \varepsilon_8^{14} \end{bmatrix}\right\}}_{\hat{\Lambda}} \otimes \begin{bmatrix} OP_1(x) \\ \vdots \\ OP_8(x) \end{bmatrix} \tag{8.6}
$$

Therefore, as shown in Eq. (8.6) and Eq. (8.1), the number of parameters to be optimized is doubled compared to DARTS. This means the time cost of optimization is twice as much in theory due to the variational inference and the differentiable search strategy.

Scale Mixture Prior

In the implementation of DARTS, each row of the sampling matrix $\boldsymbol{A}$ is constrained by the Softmax normalization, leading to the winner-takes-all evaluation. In order to overcome this phenomenon, a soft constraint based on the *MAP* is given to replace Softmax, which assumes each point of the matrix $\boldsymbol{A}$ obeys a bimodal distribution:

$$\begin{aligned} \boldsymbol{A}^{MAP} &= \arg\min_{A} -\log P(\boldsymbol{A}|D) \\ &= \arg\min_{A} -\log P(D|\boldsymbol{A}) - \log P(\boldsymbol{A}) \end{aligned} \tag{8.7}$$

Note that the parameters in Eq. (8.7) are points other than distributions. It is a coincidence that the regularization $-\log P(\boldsymbol{A})$ is an instance of the second item $-\log P((\Lambda_i^j)^{k=1})$ in Eq. (8.5) when $(\Lambda_i^j)^{k=1}$ is just the mean of the posterior distribution. Therefore, provided that the prior $P(\boldsymbol{\Lambda})$ is a bimodal distribution, this soft constraint on the sampling distributions group can be implemented in a broader sense. The prior is set to:

$$P(\boldsymbol{\Lambda}) = \prod_{j=1}^{14}\prod_{i=1}^{8} \lambda N\left(\left(\Lambda_i^j\right)^{k=1} | \mu_1, \sigma_1^2\right) + (1-\lambda) N\left(\left(\Lambda_i^j\right)^{k=1} | \mu_2, \sigma_2^2\right) \tag{8.8}$$

where $(\Lambda_i^j)^{k=1}$ is a sample point from the posterior distribution in the group, $N((\Lambda_i^j)^{k=1} | \mu, \sigma^2)$ is the Gaussian density evaluated at $(\Lambda_i^j)^{k=1}$ with mean μ and variance σ^2, λ is a hyper-parameter of the ratio scaling two Gaussian peaks. As shown in Figure 8.2, our prior is a bimodal distribution, then the group $\boldsymbol{\Lambda}$ can be divided into two peaks, and λ determines the proportion of allocation.

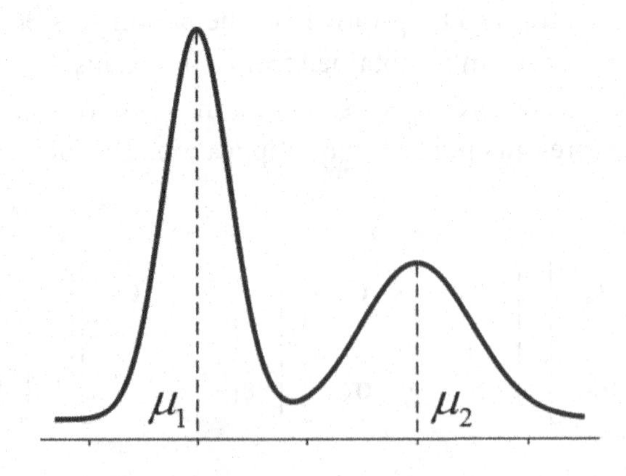

FIGURE 8.2 Bimodal prior distribution.

Forward Sampling from the Posterior Distributions Group

By optimizing the sampling distributions group, the posterior is obtained to estimate the uncertainty for model matching. Then the candidate architectures set can be sampled from the posterior via random sampling. There is no time sequence or optimization in the sampling, so it is convenient for parallel acceleration. In fact, a strategic sampling approach is allowed, e.g., the margin-focused sampling strategy, but in practice, the random sampling is sufficient to find the set. Two metrics, accuracy and sparseness, are used to evaluate the sampled architecture, where the accuracy is for the validation dataset, sparseness is for the sampling instance matrix (Hoyer 2004):

$$Sparseness(\hat{\Lambda}) = \frac{\sqrt{m} - \left(\sum_{i,j} \left|\hat{\Lambda}_i^j\right|\right) \Big/ \sqrt{\sum_{i,j} \left(\hat{\Lambda}_i^j\right)^2}}{\sqrt{m} - 1} \tag{8.9}$$

where m is the dimensionality of $\hat{\Lambda}$ and m is equal 14×8. By sorting the sparseness of the sampling instance matrix, the deviation caused by sampling from the continuous space is reduced. It is worth pointing out the difference between our method and the stochastic neural architecture search (SNAS) (Xie et al. 2018). The sampling in our work is subject to the 3σ principle as the posterior is Gaussian, and the uncertainty estimation of model matching is also given, whereas sampling in SNAS is realized by a random one-hot mask.

EXPERIMENTAL STUDIES FOR AEROENGINE BEVEL GEAR FAULT DIAGNOSIS

In this section, the aeroengine bevel gear dataset is collected and used to verify the performance of BDAS. The comparison experiments are carried out on four widely used models, i.e., the standard DCNN, VGG11, ResNet18 and Inception. The standard DCNN has a series stack of 5 convolutional layers. The VGG11, ResNet18 and Inception are directly copied from a benchmark study (Zhao et al. 2020). All the experiments are running on the Ubuntu 18.04 GNU/Linux and GeForce RTX TITAN.

Experimental Description

Data Description

The bevel gear is an important part of the aeroengine transmission system, so it is necessary to monitor its state. In the testing bench, we set up four groups of bevel gear experiments in different states, as shown in Figure 8.3, including healthy state, tooth surface wear, broken tooth and small end collapse. The bevel gears corresponding to these states were manually preprocessed and set. Each of these gears was operated under five rotational speeds, respectively, as described in Table 8.1. The vibration signals were acquired by an acceleration sensor.

For each rotational speed, the ratio of training set and test set is 2:1, and one sample has 1024 points without data normalization or augmentation. An illustration of the raw vibration signal in time domain is shown in Figure 8.4, and each segment represents a sample with 1024 points. Samples of different speeds in the same mode share the same

FIGURE 8.3 The aeroengine bevel gears in four states: (a) Healthy state; (b) tooth surface wear; (c) broken tooth; (d) small end collapse.

TABLE 8.1 The Aeroengine Bevel Gear Dataset

Data Type	Healthy State	Tooth Surface Wear	Broken Tooth	Small End Collapse
Rotational Speed		**500/1000/1500/2000/3900 rpm**		
Samples for training	600*5	600*5	600*5	600*5
Samples for testing	300*5	300*5	300*5	300*5
Label	0	1	2	3

label. Details are shown in Table 8.1. Therefore, it can be seen as a multi-condition classification problem.

Noise Addition to Simulate Multiple Data Domains

The vibration signal is often contaminated by noise in changing environment, and the data in one domain would be shifted accordingly. Therefore, to simulate data in multiple

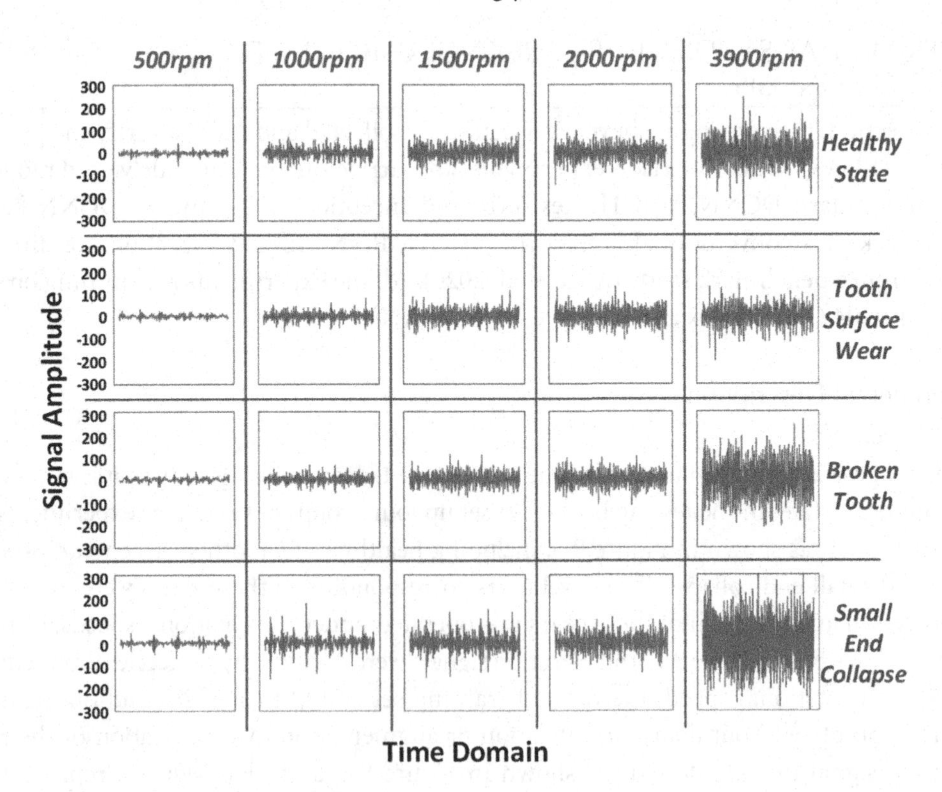

FIGURE 8.4 Illustration of the raw vibration signal in time domain.

domains problem, different levels of Gaussian white noises are added into the raw vibration signal. As the raw vibration signal inevitably contains measurement noise, we cannot obtain the noise free signal. Therefore, the raw noisy vibration signal is taken as the benchmark to generate Gaussian white noise with different energy levels. The definition of noise energy level (NEL) is similar to the signal-to-noise ratio, as described in Eq. (8.10):

$$\mathrm{NEL} = 10\log_{10}\left(\frac{P_{\mathrm{noise}}}{P_{\mathrm{signal}}}\right) \tag{8.10}$$

where P_{noise} is the noise energy, P_{signal} is the signal energy. The NELs in this study are (10, 5, 0, −5, −10) respectively. The higher is the NEL, the more serious is the contamination of the raw signal. Then, multi-domain problem can be obtained by adding different energy levels of noise.

Results and Discussion

Results

Search Space The search space includes four nonparametric operators (*zero, max-pool, avg-pool, skip-connection*) and four parametric operators (*sep-conv-3, sep-conv-5, dil-conv-3, dil-conv-5*).

Settings The setting of the proposed BDAS method has three parts: (1) In the training phase, we warm $OP_{5\text{-}8}$ up separately for 20 epochs. Then the path-dropout training is used to alleviate the co-adaptation in hyper-network for 150 epochs, and the ratio is set to 0.2. Stochastic Gradient Descent (SGD) is used to optimize the weights with learning rate 0.001. The gradient update of the masked paths is frozen in both warmup and path-dropout. (2) In the searching phase, the parameter θ of each distribution of the sampling distributions group Λ is initialized as (0.2, −4), the parameter $(\lambda, \mu_1, \mu_2, \sigma_1, \sigma_2)$ of the prior $P(\Lambda)$ is initialized as (0.8, 0.1, 0.3, 0.03, 0.05). The parameters setting meets the requirement as long as the bimodal prior distribution and the initialized posterior distribution conform to such structure in Figure 8.5. Adam is used to optimize the parameter θ with learning rate 0.001 for 30 epochs. Then 2000 architectures are sampled from the optimized posterior and retain the top 10 ordered by accuracy and sparseness as the final set. (3) In the retraining phase for the final 10 architectures, SGD is used to optimize the model for 300 epochs and the cosine annealing learning rate scheduler is used with the learning rate ranging from 0.03 to 0.002.

To get the top 10 architectures, the sampled 2000 architectures are ordered by validation accuracy and sparseness. At first, the top 50 architectures are selected as s subset by validation accuracy. After that, the top 10 are selected from the top 50 subset as the final set by sparseness. Finally, the top 10 architectures are retrained and reordered.

The results of BDAS and other comparative models are shown in Table 8.2. In order to eliminate the randomness of the results, the testing accuracy is the average of the last 10 epochs, and the value in parenthesis is the variance of accuracy. It can be seen that BDAS achieves the upper bound compared with other comparative models and the baseline

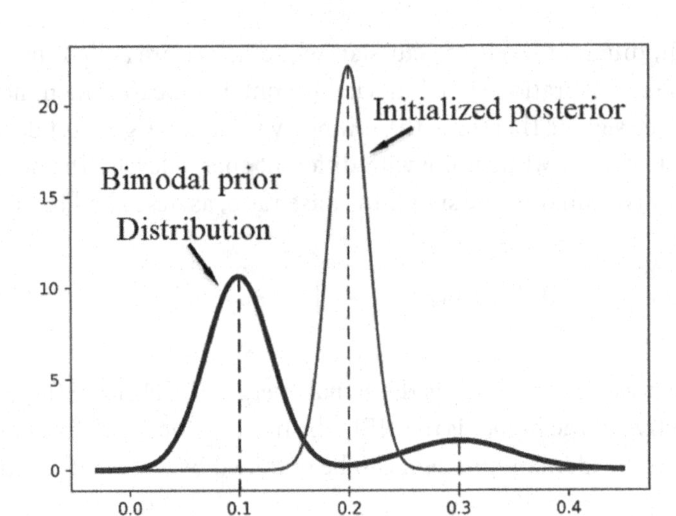

FIGURE 8.5 Bimodal prior distribution and the initialized posterior.

DARTS in the simulated multiple data domains. Especially, BDAS outperforms DCNN up to 19.09% and outperforms DARTS up to 4.44% in the 10 NEL case. Besides, BDAS is very competitive in model size compared with manually designed networks, and the lightweight model has superiority when applied to the resource-constrained scenarios. Figure 8.6 shows the degradation of model performance with the increase of noise. The accuracy of the standard DCNN decreases the most, while that of BDAS decreases the least.

In fact, BDAS gets ten architectures during one searching phase. Figure 8.7 shows the box plots of these sets for corresponding noise energy levels. It can be seen that the performance of the architecture searched by DARTS is in the distribution of the set searched by BDAS. The lower the green square is in the boxplot, the more architectures BDAS finds that exceed DARTS. This not only confirms the hypothesis in (Zhang et al. 2020), but also shows the effectiveness of our method. By applying the variational inference into the differentiable search strategy, it is efficient to get a set with several high performance architectures.

Discussion of Warmup and Path Dropout

To verify the effectiveness of our training strategy for the hyper-network, the ablation study is carried out in the case of 0 NEL. In the proposed BDAS method, warmup is used

TABLE 8.2 The Experimental Results

Model Name	Model Size	Testing Accuracy under five NELs					Testing Accuracy for Raw Signal
		10	5	0	−5	−10	
DCNN	6.4M	36.20 (0.15)	62.79 (0.07)	87.02 (0.01)	96.16 (0.01)	99.09 (0)	99.67 (0)
VGG11	31.3M	44.49 (0.09)	65.74 (0.06)	88.02 (0.02)	96.53 (0.1)	99 (0.01)	99.59 (0.01)
ResNet18	15.4M	44.37 (0.13)	65.28 (0.04)	86.08 (0.05)	94.99 (0.02)	98.11 (0.01)	99.37 (0.01)
Inception	**3.3M**	48.35 (0.12)	64.72 (0.08)	82.21 (0.03)	92.76 (003)	96.58 (0.02)	99.22 (0)
DARTS	6.3M	50.85 (0.65)	69.69 (0.15)	90.32 (0.27)	97.19 (0.04)	99.49 (0)	99.71 (0.03)
BDAS	6.3M	**55.29** (2.7)	**70.12** (0.23)	**90.91** (0.2)	**98.11** (0.02)	**99.57** (0)	**99.89** (0)

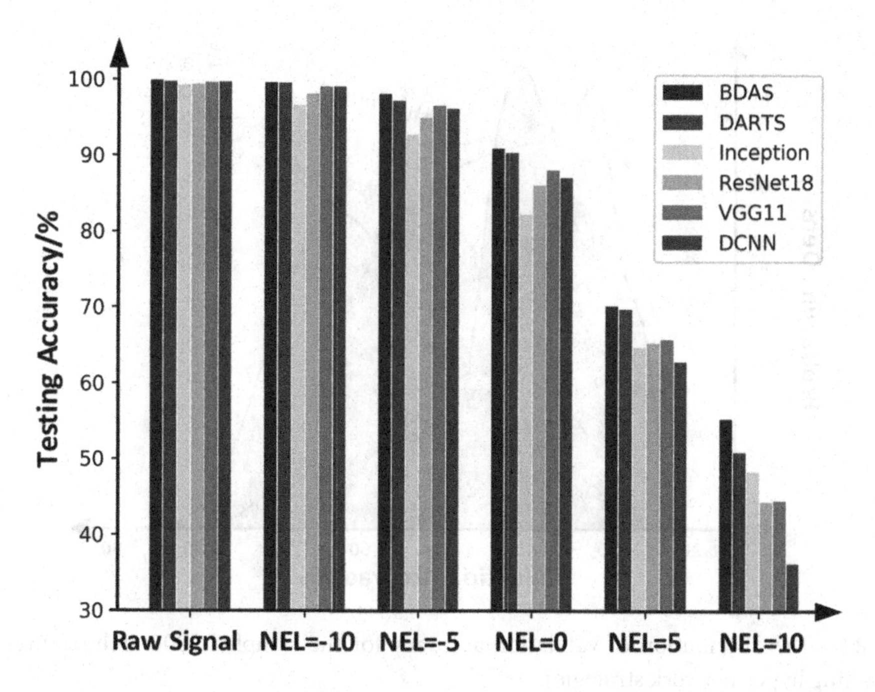

FIGURE 8.6 The degradation of model performance with the increase of noise.

to enhance the competitiveness of the parametric operators and path-dropout is used to alleviate the co-adaptation problem. In the One-shot method, only path-dropout is used to train the hyper-network. The baseline DARTS takes no strategy in the pre-training phase. These three strategies are respectively used to train the hyper-network for basal 150 epochs, among which the dropout ratio of both BDAS and One-shot is set to 0.2. Especially, BDAS will warm up the $OP_{5\text{-}8}$ separately for 20 epochs, and One-shot and DARTS will train the

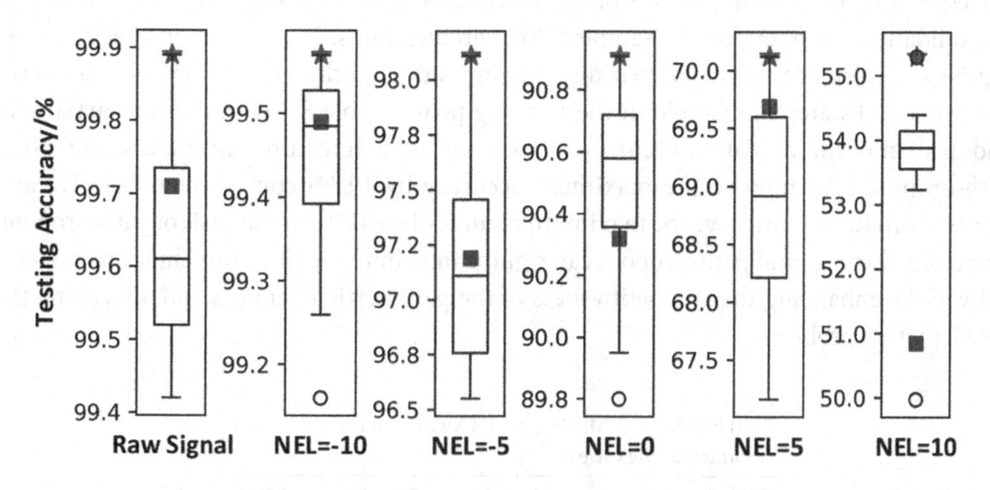

FIGURE 8.7 The boxplots of the testing accuracy of the set searched by BDAS. The hollow circles represent the outliers of the boxplots. The red pentacles are the result searched by BDAS. The green squares represent the result of DARTS.

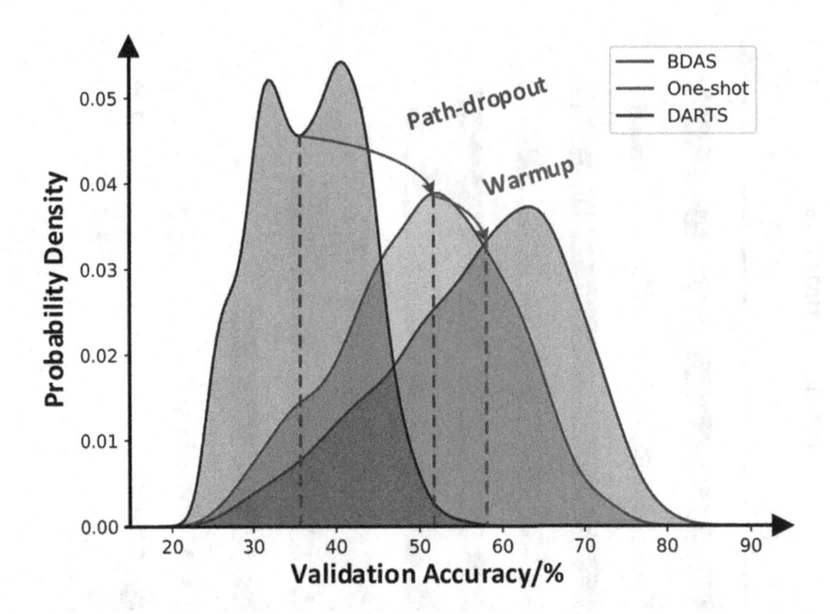

FIGURE 8.8 The distribution of validation accuracy for the sampled 2000 architectures under three training hyper-network strategies.

hyper-network with 20 more epochs for the sake of fairness. Then we get the corresponding three hyper-networks, and their training accuracy and validation accuracy are all about 98% and 80% respectively. It seems that the representation of these three hyper-networks remains the same level. However, the subsequent experiment shows that there are indeed significant differences between them.

After the training phase, the same Bayesian differentiable search strategy is used to optimize the sampling distributions group, and 2000 architectures are sampled, respectively, corresponding to three hyper-networks. Figure 8.8 shows the kernel density estimation of the validation accuracy for the sampled 2000 architectures. It can be seen that there are significant differences between their distributions although the accuracy levels of the three hyper-networks are comparable in the training phase. Table 8.3 describes the maximum and median of the validation accuracy of the sampled architectures. In the case of 0 NEL, path-dropout can improve the maximum accuracy by 19.3% compared to DARTS, and warmup can further improve the maximum accuracy by 6.08% on the basis of path-dropout. Therefore, warmup and path-dropout can significantly improve the representation of hyper-network by enhancing the competitiveness of the parametric operators and alleviating the co-adaptation problem.

TABLE 8.3 The Maximum and Median of the Validation Accuracy in the Case of 0 NEL

Validation Accuracy	DARTS	One-Shot	BDAS
Median	36.78	51.32	**58.63**
Maximum	55.75	75.05	**81.13**

Discussion of Sparseness

In practice, it has been found that the sampled architecture with the maximum validation accuracy was not the best model in the testing phase (Li et al. 2020). The reason for this phenomenon lies in the bias between the hyper-network and the sub-network, which is the inherent problem of the differentiable search strategy. As mentioned above, DARTS will select the operators with the maximum coefficients from each of the mixed edges to construct the final searched sub-network. The other seven operators in one mixed edge will be dropped, although they also provide feature in the hyper-network as their coefficients are not zero. It is the deletion that results in the bias between the hyper-network and sub-network. BDAS not only inherits the high efficiency of the differentiable search strategy, but also suffers from the bias problem. Therefore, to alleviate this problem, sparseness is used to assist the validation accuracy to select sub-networks. Firstly, we select the architectures with top 50 validation accuracy as a subset. Then the architectures with top 10 sparseness are selected from the subset as the final set. Figure 8.9 shows the bivariate distribution of validation accuracy and sparseness for the top 50 architectures of each noise level. Table 8.4 shows the testing accuracy and sparseness of the architecture with the highest validation accuracy and the best architecture in the set with top 10 sparseness. It can be seen that the architectures with the highest validation accuracy are not the best models in the testing phase. An architecture balancing validation accuracy and sparseness is more likely to perform better

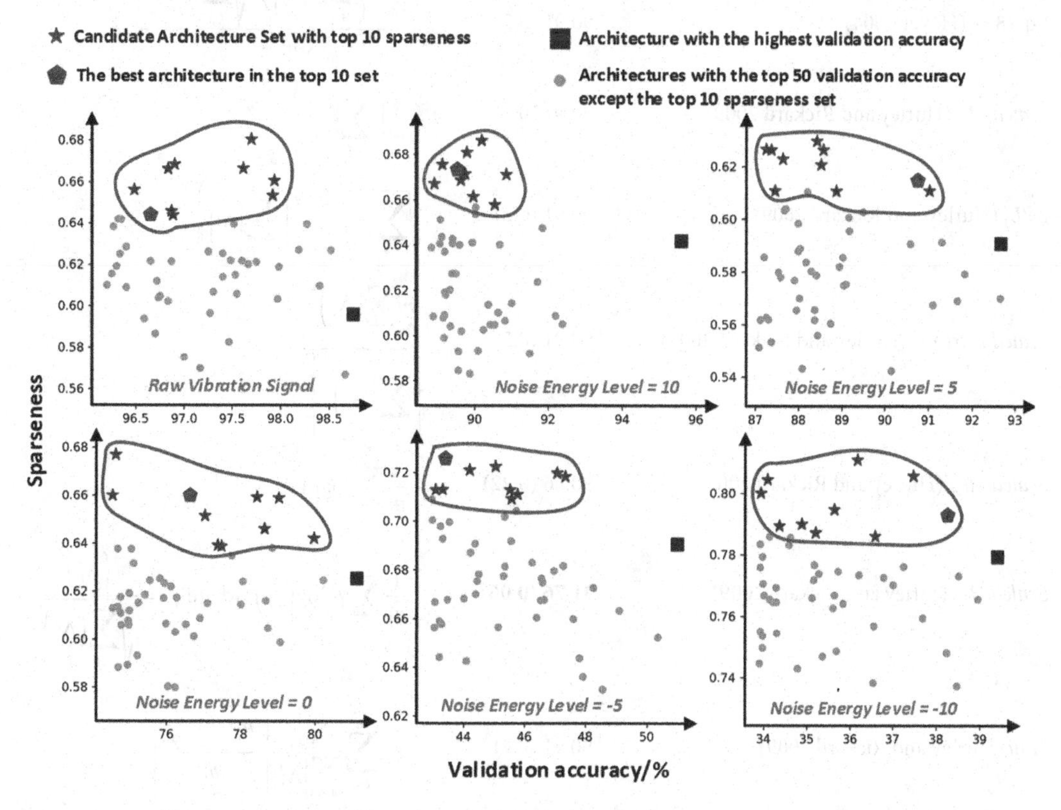

FIGURE 8.9 The scatterplots of bivariate distribution of validation accuracy and sparseness for each data domain.

TABLE 8.4 Comparison of the Architecture with the Highest Validation Accuracy and the Best Architecture in the Set with Top 10 Sparseness

	NEL = 10		NEL = 5		NEL = 0		NEL = −5		NEL = −10		Raw Signal	
	Net-A	**Net-S**	**Net-A**	**Net-S**	**Net-A**	**Net-S**	**Net-A**	**Net-S**	**Net-A**	**Net-S**	**Net-A**	**Net-S**
Sparseness	0.78	0.79	0.69	0.73	0.63	0.66	0.59	0.61	0.64	0.67	0.60	0.64
Testing accuracy	55.01	**55.29**	68.41	**70.12**	90.41	**90.91**	97.54	**98.11**	99.42	**99.57**	99.25	**99.89**

Net-A is the architecture with the highest validation accuracy. Net-S is the best architecture in the set with top 10 sparseness.

in the testing phase. Therefore, the sparseness metric can help BDAS find the best one in the architectures set.

We carried out a comparative experiment to further evaluate the influence of sparsity measure. In the case of 0 NEL, we chose several sparsity measures from (Hurley and Rickard 2009) to replace Eq. (8.9), and the other experimental settings were the same as aforementioned. As shown in Table 8.5, the *Scaled Hs* measure can indicate a better

TABLE 8.5 The Comparative Experiment of Sparsity Measures in the Case of 0 NEL

Sparsity Measure	Testing Accuracy	Definition
Eq. (8.9) (Hoyer 2004)	90.91 (0.2)	$\frac{\sqrt{m}-\left(\sum_{i,j}\left\lvert\hat{\Lambda}_i^j\right\rvert\right)\Big/\sqrt{\sum_{i,j}\left(\hat{\Lambda}_i^j\right)^2}}{\sqrt{m}-1}$
Scaled $-L_1$ (Hurley and Rickard 2009)	90.91 (0.2)	$-\frac{1}{m}\left(\sum_{i,j}\left\lvert\hat{\Lambda}_i^j\right\rvert\right)$
L_2/L_1 (Hurley and Rickard 2009)	90.91 (0.2)	$\sqrt{\sum_{i,j}\left(\hat{\Lambda}_i^j\right)^2}\Big/\left(\sum_{i,j}\left\lvert\hat{\Lambda}_i^j\right\rvert\right)$
Scaled Kurtosis (Hurley and Rickard 2009)	90.91 (0.2)	$\frac{m\sum_{i,j}\left(\hat{\Lambda}_i^j\right)^4}{\left[\sum_{i,j}\left(\hat{\Lambda}_i^j\right)^2\right]^2}$
Scaled –log (Hurley and Rickard 2009)	90.66 (0.22)	$-\frac{1}{m}\sum_{i,j}\log\left[1+\left(\hat{\Lambda}_i^j\right)^2\right]$
Scaled Hs (Hurley and Rickard 2009)	**91.76 (0.06)**	$-\frac{1}{m}\sum_{i,j}\gamma_i^j\log\left(\gamma_i^j\right)^2$ where $\gamma_i^j=\frac{\left(\hat{\Lambda}_i^j\right)^2}{\sqrt{\sum_{i,j}\left(\hat{\Lambda}_i^j\right)^2}}$
Gini (Hurley and Rickard 2009)	90.91 (0.2)	$1\text{-}2\sum_{k=1}^{m}\frac{\Lambda_{(k)}}{\sum_{i,j}\left\lvert\hat{\Lambda}_i^j\right\rvert}\left(\frac{m-k+\frac{1}{2}}{m}\right)$ for ordered data, $\Lambda_{(1)}\le\Lambda_{(2)}\le\cdots\le\Lambda_{(m)}$

model and the *Scaled –log* measure was failed to find an equivalent model compared with other five measures in this case. Due to the inevitable deviation between the sampled sub-network and the hyper-network, we need to consider both the validation accuracy and sparsity measure when evaluating the results, as the sparsity measure can be used to quantify this deviation.

CONCLUSION

In this work, we have proposed a one-shot NAS method to automatically design domain matching fault diagnostic model. As a rethinking extension, our method solves two drawbacks of DARTS, i.e., the unfairness of hyper-network and the overconfident result. The warmup and path-dropout are used to obtain an intuitively fair hyper-network, and variational Bayesian inference is introduced into the differentiable search strategy to estimate the uncertainty of model matching and get the set with multiple architectures. Our approach achieves state-of-the-art performance on the aeroengine bevel gear fault diagnosis with different NEL cases.

REFERENCES

Baruwa, Ahmed, Mojeed Abisiga, Ibrahim Gbadegesin, and Afeez Fakunle. 2019. "Leveraging End-to-End Speech Recognition with Neural Architecture Search." arXiv:1912.05946. Accessed December 01, 2019. https://doi.org/10.48550/arXiv.1912.05946.

Bender, Gabriel, Pieter-Jan Kindermans, Barret Zoph, Vijay K. Vasudevan, and Quoc V. Le. 2018. "Understanding and Simplifying One-Shot Architecture Search." International Conference on Machine Learning, July.

Blundell, Charles, Julien Cornebise, Koray Kavukcuoglu, and Daan Wierstra. 2015. *Weight Uncertainty in Neural Networks*. Machine Learning, arXiv: Machine Learning, May.

Chu, Xiangxiang, Bo Zhang, and Ruijun Xu. 2021. "FairNAS: Rethinking Evaluation Fairness of Weight Sharing Neural Architecture Search." 2021 IEEE/CVF International Conference on Computer Vision (ICCV). https://doi.org/10.1109/iccv48922.2021.01202.

Kingma, P. Diederik, and Max Welling. 2013. "Auto-Encoding Variational Bayes." arXiv:1312.6114. Accessed December 01, 2013. https://doi.org/10.48550/arXiv.1312.6114.

Guo, Zichao, Xiangyu Zhang, Haoyuan Mu, Wen Heng, Zechun Liu, Yichen Wei, and Jian Sun. 2019. "Single Path One-Shot Neural Architecture Search with Uniform Sampling." arXiv:1904.00420. Accessed March 01, 2019. https://doi.org/10.48550/arXiv.1904.00420.

Hoyer, PatrikO. 2004. "Non-Negative Matrix Factorization with Sparseness Constraints." arXiv: Learning,arXiv: Learning, August.

Hurley, Niall P., and Scott T. Rickard. 2009. "Comparing Measures of Sparsity." *IEEE Transactions on Information Theory* 55 (10): 4723–4741. https://doi.org/10.1109/Tit.2009.2027527.

Hu, Zhong-Xu, Yan Wang, Ming-Feng Ge, and Jie Liu. 2020. "Data-Driven Fault Diagnosis Method Based on Compressed Sensing and Improved Multiscale Network." *IEEE Transactions on Industrial Electronics* 67: 3216–25. https://doi.org/10.1109/tie.2019.2912763.

Jiang, Guoqian, Haibo He, Ping Xie, and Yufei Tang. 2017. "Stacked Multilevel-Denoising Autoencoders: A New Representation Learning Approach for Wind Turbine Gearbox Fault Diagnosis." *IEEE Transactions on Instrumentation and Measurement*, 66: 2391–2402. https://doi.org/10.1109/tim.2017.2698738.

Klyuchnikov, Nikita, Ilya Trofimov, Ekaterina Artemova, Mikhail Salnikov, MaximV. Fedorov, and Evgeny Burnaev. 2020. "NAS-Bench-NLP: Neural Architecture Search Benchmark for Natural Language Processing." *IEEE Access* 10. https://doi.org/10.1109/Access.2022.3169897. <Go to ISI>://WOS:000790760000001.

Liang, Hanwen, Shifeng Zhang, Jiacheng Sun, Xingqiu He, Weiran Huang, Kechen Zhuang, and Zhenguo Li. 2019. DARTS+: Improved Differentiable Architecture Search with Early Stopping. arXiv:1909.06035. Accessed September 01, 2019. https://doi.org/10.48550/arXiv.1909.06035.

Li, Guohao, Guocheng Qian, Itzel C. Delgadillo, Matthias Muller, Ali Thabet, and Bernard Ghanem. 2020. "SGAS: Sequential Greedy Architecture Search." In 2020 IEEE/CVF Conference on Computer Vision and Pattern Recognition (CVPR). https://doi.org/10.1109/cvpr42600.2020.00169.

Liu, Hanxiao, Karen Simonyan, and Yiming Yang. 2018. "DARTS: Differentiable Architecture Search." arXiv:1806.09055. Accessed June 01, 2018. https://doi.org/10.48550/arXiv.1806.09055.

Li, Tianfu, Zhibin Zhao, Chuang Sun, Ruqiang Yan, and Xuefeng Chen. 2020. "Adaptive Channel Weighted CNN With Multisensor Fusion for Condition Monitoring of Helicopter Transmission System." *IEEE Sensors Journal* 20 (15): 8364–73. https://doi.org/10.1109/jsen.2020.2980596.

Li, Tianfu, Zhibin Zhao, Chuang Sun, Li Cheng, Xuefeng Chen, Ruqiang Yan, and Robert X. Gao. 2022. "WaveletKernelNet: An Interpretable Deep Neural Network for Industrial Intelligent Diagnosis." *IEEE Transactions on Systems, Man, and Cybernetics: Systems* 52: 2302–12. https://doi.org/10.1109/tsmc.2020.3048950.

Real, Esteban, Alok Aggarwal, Yanping Huang, and Quoc V. Le. 2019. "Regularized Evolution for Image Classifier Architecture Search." Proceedings of the AAAI Conference on Artificial Intelligence, September, 4780–89. https://doi.org/10.1609/aaai.v33i01.33014780.

Sun, Jiedi, Changhong Yan, and Jiangtao Wen. 2018. "Intelligent Bearing Fault Diagnosis Method Combining Compressed Data Acquisition and Deep Learning." *IEEE Transactions on Instrumentation and Measurement* 67: 185–95. https://doi.org/10.1109/tim.2017.2759418.

Wang, Ruixin, Hongkai Jiang, Xingqiu Li, and Shaowei Liu. 2020. "A Reinforcement Neural Architecture Search Method for Rolling Bearing Fault Diagnosis." *Measurement* 154 (March): 107417. https://doi.org/10.1016/j.measurement.2019.107417.

Wang, Hao, and Dit-Yan Yeung. 2016. "Towards Bayesian Deep Learning: A Framework and Some Existing Methods." *IEEE Transactions on Knowledge and Data Engineering* 28: 3395–3408. https://doi.org/10.1109/tkde.2016.2606428.

Xie, Sirui, Hehui Zheng, Chunxiao Liu, and Liang Lin. 2018. "SNAS: Stochastic Neural Architecture Search." arXiv:1812.09926. Accessed December 01, 2018. https://doi.org/10.48550/arXiv.1812.09926.

Xu, Hang, Lewei Yao, Zhenguo Li, Xiaodan Liang, and Wei Zhang. 2019. "Auto-FPN: Automatic Network Architecture Adaptation for Object Detection Beyond Classification." 2019 IEEE/CVF International Conference on Computer Vision (ICCV). https://doi.org/10.1109/iccv.2019.00675.

Zhang, Yuge, Quanlu Zhang, and Yaming Yang. 2020. "How Does Supernet Help in Neural Architecture Search?" arXiv:2010.08219. Accessed October 01, 2020. https://doi.org/10.48550/arXiv.2010.08219.

Zhao, Zhibin, Tianfu Li, Jingyao Wu, Chuang Sun, Shibin Wang, Ruqiang Yan, and Xuefeng Chen. 2020. "Deep Learning Algorithms for Rotating Machinery Intelligent Diagnosis: An Open Source Benchmark Study." *ISA Transactions* 107: 224–55. https://doi.org/10.1016/j.isatra.2020.08.010.

Zoph, Barret, and Quoc V. Le. 2016. "Neural Architecture Search with Reinforcement Learning." arXiv:1611.01578. Accessed November 01, 2016. https://doi.org/10.48550/arXiv.1611.01578.

CHAPTER 9

Self-Supervised Learning for Intelligent Fault Diagnosis

INTRODUCTION

With the advancement of industrial modernization, the designs of massive large-scale mechanical equipment such as helicopters, wind turbines, and aeroengines are becoming more and more complicated, which increases instability of the mechanical systems. If a sudden failure occurs, it might result in significant downtown loss and severe catastrophic accidents. Therefore, timely and effective diagnosis of various fault states is of vital importance to avoid economic loss and improve the safety of equipment (Jin et al. 2013; Deutsch and He 2017).

In the existing research, machinery fault diagnosis methods can be broadly categorized into signal processing methods (Choi et al. 2010) and machine learning methods (Sun et al. 2018). Signal processing methods aim to extract fault-related frequency components from recorded signals. However, it is challenging to deal with complex signals when the characteristic frequency is submerged in substantial noise. Traditional machine learning methods extract hand-crafted features via exquisite structural design, and then construct the mapping between the features and the mechanical health status in a data-driven manner. Though they can achieve good performance, the procedure of feature extraction still relies on expert knowledge. As a sub-field of machine learning, deep learning (DL) can match the high nonlinear relationship between input and output without any data preprocessing and achieve higher diagnostic accuracy. Thus, it has been widely applied in fault diagnosis (Wu et al. 2021; Zhao et al. 2020).

Nevertheless, most DL-based techniques follow a supervised learning paradigm, which requires enough labeled samples to guarantee the steady convergence of the model. However, the amount of labeled data is insufficient in real-world applications. On the one hand, the occurrence of failures in key components is scarce and the expense of fault simulation experiments is high. On the other hand, unlike perceptual data such as image and acoustic signals, marking the mechanical fault type depends heavily on domain

DOI: 10.1201/9781003474463-11

expert knowledge, which is time-consuming. Unfortunately, when only few labeled data are available, the deep model will fall into the dilemma of overfitting and poor generalization ability. Therefore, it is urgent to address the label-insufficient problem in the mechanical industry.

To overcome the above problem, an intuitive idea is to enrich the capacity of labeled samples. Data augmentation (DA) (Zhuo and Ge 2021) is a simple yet effective approach to increase the number of label-invariant data via different transformations. Nonetheless, traditional DA methods such as rotation, crop, or some of their combinations were initially designed for images and are hence inapplicable to vibration data. In recent years, researchers have adopted transfer learning as a solution to the label-insufficient problem, which attempts to transfer the general representation learned from source data with sufficient labeled samples to target data with scarce label information. It performs well on limited target data, but enough labeled source data is the prerequisite of transfer learning, the label-insufficient problem has not been radically solved. Semi-supervised learning (SSL) (Zhao et al. 2016; Luo et al. 2017; Razavi-Far et al. 2018; Chen et al. 2019; Li, Li, and Ma 2020; Nie and Xie 2021) is a popular and effective method to deal with label scarcity which integrates few labeled samples with massive unlabeled samples to improve the discriminability of the model. For example, Nie and Xie (2021) pre-trained a convolutional gated recurrent unit (ConvGRU) network to classify the health conditions and later adopted three regularization terms to revise the initial labels. Li, Li, and Ma (2020) adopted an autoencoder for feature extraction and created pseudo-labels for unlabeled samples by distance metric learning. Luo et al. (2017) integrated an adaptive similarity matrix learning method into the semi-supervised feature selection procedure which achieved good performance for video semantic recognition with few labeled instances. However, most SSL methods learn representations based on clustering or manifold assumption, which requires cumbersome feature selection and supervised fine-tuning. Besides, the performance is highly related to the quality of learned representations.

As a powerful unsupervised representation learning method, self-supervised learning can extract useful representations from unannotated data. Therefore, it is an intuitive choice to combine self-supervised learning in a unified SSL framework, which avoids complex feature selection and leverages the strong feature extraction ability of self-supervised learning to enhance the model's discriminability with only few labeled data. The underlying idea of self-supervised learning is to learn invariant representations under various data distortions. Specifically, different augmentations of raw data are projected into an embedding space through Siamese networks (Koch, Zemel, and Salakhutdinov 2015; Sung et al. 2018), where the similarity of representations generated from similar samples is maximized. Following this pipeline, several contrastive learning works (Dosovitskiy et al. 2014; Chen et al. 2020; Falcon and Cho 2020; He et al. 2020) consider each sample as a class and generate different instances to distinguish whether they are from the same class. Though these methods have achieved good performance, they only focus on the instance-level invariant features regardless of the data structure itself. Recent studies have shown that an instance can be more informative when its contextual structure relationship is considered. For example, CPC (Oord, Li, and Vinyals 2018) extracted unsupervised features for speech recognition by maximizing the similarity between audio and its context segments. Particularly for fault diagnosis, experts tend to identify the fault types according to the varying tendency of vibration signals, which reveals that the temporal relation contains related information on fault characteristics. Intuitively, combining both inter-sample correlation and intra-sample temporal dependency can further increase the model's discriminability. Therefore, besides the supervised learning with labeled data, our work aims to learn representations from the perspective of inter-instance and intra-temporal relations to fully exploit the intrinsic features of unlabeled vibration data.

Based on the above discussion, we propose a SSL framework, where supervised and self-supervised tasks are jointly trained in an end-to-end manner. Besides, both inter-sample and intra-temporal relations of vibration signals are explored to enhance the discriminability. To be specific, we first apply a specific signal augmentation technique that increases the number of labeled samples and generates correlated domains of the input samples for contrastive learning. Then a supervised learning task and two self-supervised contrasting tasks are aggregated in this SSL framework, which can extract more discriminative features via designed inter-instance and intra-temporal relation mining modules. Moreover, to avoid the single task domination problem during the training procedure, a dynamic weighting mechanism is leveraged to dynamically balance the weights of each task.

To implement fault diagnosis with few labeled data, this chapter merges self-supervised learning to a semi-supervised framework, where massive unannotated samples and few labeled data are parallelly processed and fully exploited in an end-to-end manner. The framework provides a synthetical consideration of two types of relations, including inter-sample and intra-temporal relations of vibration signals, which further enhance the overall diagnostic performance. Besides, a data augmentation method of vibration signals is pertinently designed to improve the generalization ability of the model and create different domains for self-supervised learning. Finally, a dynamic weighting mechanism is adopted to assign the weights of multi-task losses automatically for better optimization.

The rest of this chapter is organized as follows: background and definition of SSL and self-supervised learning, SSL-based intelligent fault diagnosis framework, datasets, evaluation results, and further discussions.

THEORETICAL BACKGROUND

This section is going to provide background and definition of SSL and self-supervised learning.

Semi-Supervised Learning

Suppose that the collected signals consist of limited labeled dataset $D_L = \{(x_i, y_i) \mid x_i \in R^{1\times L}\}_{i=1}^{N_l}$ and massive unlabeled dataset $D_U = \{x_i \mid x_i \in R^{1\times L}\}_{i=1}^{N_u}$, where x_i is an input sample with L points and $y_i \in \{1,2,\ldots,C\}$ is its corresponding label indicating C different health conditions. N_l and N_u refer to the size of labeled dataset and unlabeled dataset respectively, which satisfy the inequation: $N_l \ll N_u$. The total number of input samples N is the sum of N_l and N_u. Let

α denote the ratio of N_l to N, then the above case can be viewed as an SSL problem with a label rate α.

In general, SSL methods are categorized into three classes. The first one is unsupervised pre-training, which initially learns representation from unlabeled data via unsupervised approaches and fine-tunes the model on labeled data. Another method is self-training (Lee 2013; Tai, Bailis, and Valiant 2021), which first trains the network on few labeled data and then utilizes the trained model to classify unlabeled samples. Afterward, samples with high classification confidence are added to the training set to re-train the network, and so on. Co-training methods (Jawed, Grabocka, and Schmidt-Thieme 2020; Laine and Aila 2016; D. Rasmus et al. 2015; Lee et al. 2022) jointly train the model with both labeled and unlabeled data, which accordingly output a supervised loss L_s and an unsupervised loss L_u. The final loss L is the weighted sum of the supervised and unsupervised loss, as formulated in:

$$L = L_s + \omega L_u \tag{9.1}$$

where ω is a hyper-parameter to adjust the weight of L_s and L_u.

To integrate self-supervised tasks into SSL, our work follows the co-training framework, which learns an end-to-end diagnostic model with few labeled data.

Self-Supervised Learning

Previous self-supervised learning works (Doersch, Gupta, and Efros 2015; Gidaris, Singh, and Komodakis 2018) use handcrafted prediction tasks to learn context-level representation, which gains discriminative information via comparing a sample with its contextual environment. For example, Doersch et al. (2015) generate different sub-segments by cutting the images and trains the model to predict their relative position. Recently, instance-instance contrastive directly learns the instance-level relationship between different samples by metric learning. Deep cluster (Caron et al. 2018) first clusters data into various classes, and then utilizes the class index of each instance as its target label. However, it is time-consuming for two-stage learning. Instance discrimination methods (Chen et al. 2020; Grill et al. 2020; He et al. 2020) propose to generate a triplet including anchor, positive and negative representations in the embedding space, where the anchor should be pulled close to the positive feature and away from the negative one. To avoid model collapse, MoCo (He et al. 2020) introduced memory bank and momentum encoder to increase more negative pairs and dynamically update the encoder. SimCLR (Chen et al. 2020) added a nonlinear mapping behind the encoder and further increased the number of negative samples. BYOL (Grill et al. 2020) proposed a self-supervised paradigm needless of negative samples. They adopted an asymmetric network, which attached a predictor to one branch of the Siamese network and stopped the gradient back-propagation of the other branch. More recently, feature decorrelation-based methods such as Barlow Twins (Zbontar et al. 2021) proposed a regularization term via decreasing the redundancy of the cross-correlation matrix, which provided a new solution for self-supervised learning.

The above methods have made a good performance. However, they neglected the influence of context information on classification tasks. In this work, we take the temporal relation of vibration signals into consideration and design two parallelized branches to capture both instance-level and temporal-level features, which improves the discriminability.

SSL-BASED INTELLIGENT FAULT DIAGNOSIS

This section is going to first introduce the overall framework of SSL-based intelligent fault diagnosis, and then demonstrate some of its specific modules, i.e., time-amplitude signal augmentation module, supervised learning module, self-supervised proxy task and uncertainty-based dynamic weighting module.

Framework for Intelligent Fault Diagnosis

The overall framework of ITSSL is divided into the training stage and the testing stage, as illustrated in Figure 9.1. During the training stage, all training samples are first transformed into different forms using a time-amplitude data augmentation technique. After that, labeled samples are fed into a backbone network to extract discriminative features and further through a classification head to obtain a supervised loss Ls. Meanwhile, positive-negative pairs are constructed from unlabeled samples by two relation mining modules respectively, which serve as the pseudo labels of self-supervised tasks. The sample pairs are then fed into the shared backbone network and passed through a corresponding output head to determine whether they belong to a positive pair or not, and output a contrastive loss, i.e., inter-instance sample pairs are fed into the instance relation head to calculate the instance relation loss Li, and intra-temporal sample pairs are fed into the temporal relation head to calculate the temporal relation loss Lt. Later, a dynamic weighted loss aggregates the supervised loss Ls, instance relation loss Li and temporal relation loss Lt simultaneously, where the weight of each loss is determined by its uncertainty. Finally, the whole model is trained with the total loss, and weights are updated by back-propagation.

In the testing stage, the instance relation head and temporal relation head are abandoned, while the testing samples are directly fed into the trained backbone network and the classification head to predict the real fault types.

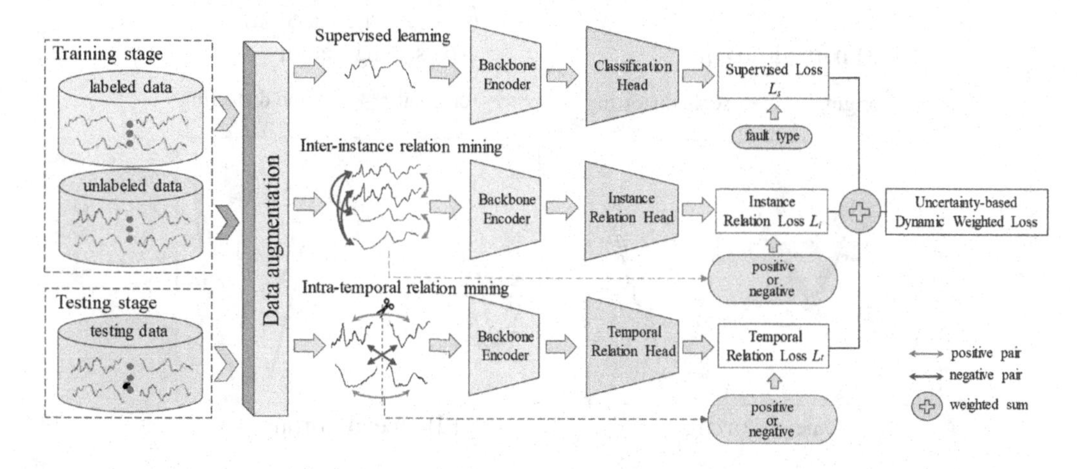

FIGURE 9.1 The overall framework of ITSSL.

Time-Amplitude Signal Augmentation

At the beginning of ITSSL, data augmentation is applied to all training samples through a series of transformations. It should be emphasized that data augmentation plays different roles for labeled samples and unlabeled samples. For labeled samples, it increases the number of label-invariant data which improves the robustness of the model. As for unlabeled samples, data augmentation serves as the foundation for the subsequent construction of positive and negative pairs. However, vibration signal augmentation techniques are scarce compared with image augmentation methods. Thus, we design a specific signal augmentation method considering vibration signal properties.

In industrial scenarios, complex noises may modulate the amplitude and frequency of the real vibration signals, which warps signals both in the time axis and amplitude axis. Motivated by this, a time amplitude distorting (TAD) technique is adopted to imitate the noise interference, as shown in Figure 9.2. First, all input samples are normalized, limiting the value of each point to the range of [−1, 1]. Then two transformation techniques, including scale distorting and time distorting, are applied in amplitude and time domains respectively. Each transformation has a 20% chance of being implemented. Scale distorting simulates the amplitude fluctuation of signals. The time steps remain constant, but the values scale with a dynamic scaling factor at each time step. As shown in Figure 9.2(b), the scaling factor follows a smooth cubic spline curve controlled by 6 equally spaced points generated from a Gaussian distribution, which enhances the robustness of the model to noise amplitude modulation.

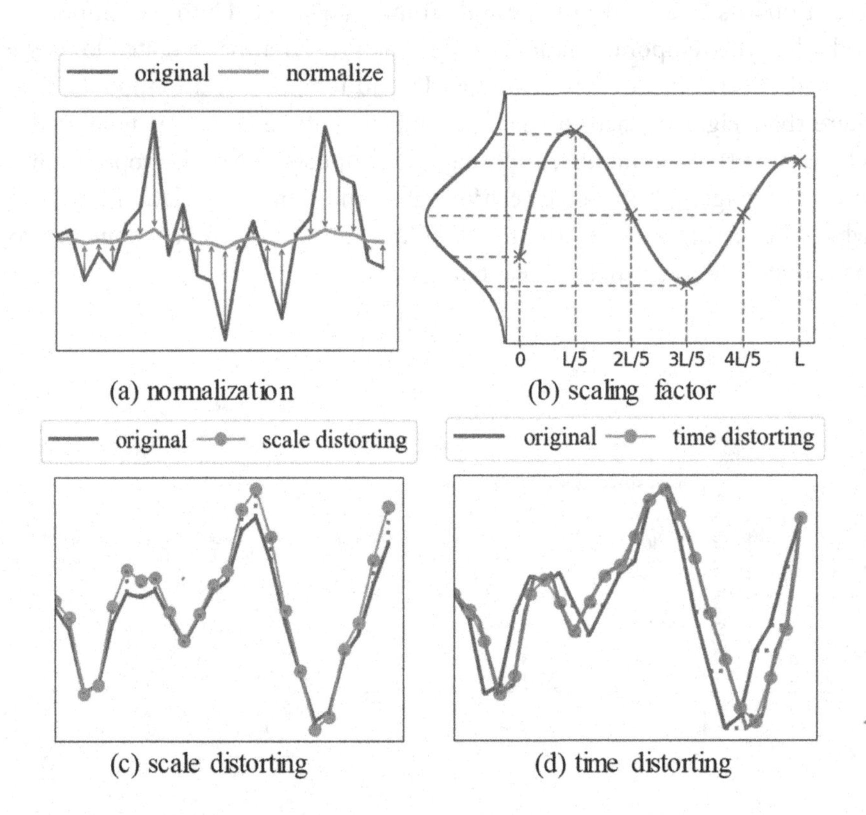

FIGURE 9.2 The time-amplitude signal augmentation technique in ITSSL.

After that, time distorting is utilized to mimic the frequency fluctuation, causing each point to shift with a time scaling factor. Similarly, the factor follows a cubic spline curve, and a new set of points are generated. The new point set is then used to interpolate the value at the original time steps and the interpolation result is the final output of time distorting. Through time distorting, the model can be more robust to signal frequency fluctuations.

Supervised Learning

Given the insufficient labeled dataset $D_L = \{(x_i, y_i) | x_i \in R^{1\times L}\}_{i=1}^{N_l}$, a backbone encoder $E_\theta(\bullet)$ takes each sample as input to extract its high-level semantic feature. After that, the hidden features are fed into a classification head $H_\varphi(\bullet)$ and output a probability distribution vector $p_i^{sup} = H_\varphi(z_i)$, which indicates the predicted health state. Finally, the prediction result p_i^{sup} and label y_i are combined to compute the cross-entropy (CE) loss L_s, which is formulated as:

$$L_s = -\frac{1}{N_l}\sum_{i=1}^{N_l} y_i \bullet \log\left(p_i^{sup}\right) \tag{9.2}$$

Self-Supervised Learning

Inter-Instance Relation Mining

The inter-instance relation mining module aims to discover the relationship between different samples. It follows the pipeline of instance discrimination-based self-supervised methods in feature extraction. Formally, for a batch of samples $\{x_i | x_i \in R^{1\times L}\}_{i=1}^{B}$ where B denotes the batch size, each sample is transformed M times by aforementioned TAD, which generates a new set of signal instances $\{x_{i,j} | x_{i,j} \in R^{1\times L}, i = 1, 2, \ldots, B, j = 1, 2, \ldots, M\}$. Based on the fact that instances generated from the same sample are similar while those from different samples differ markedly, given an anchor instance $x_{s,t}$, those instances $x_{s,t}^{+} \in \{x_{i,j} | x_{i,j} \in R^{1\times L}, i = s, j = 1, 2, \ldots, M\}$ converted from x_s are considered as its positive pairs, and other instances $x_{s,t}^{-} \in \{x_{i,j} | x_{i,j} \in R^{1\times L}, i \neq s, j = 1, 2, \ldots, M\}$ are its negative pairs, as shown in Figure 9.3(a). Then all instances are fed in the backbone encoder $E_\theta(\bullet)$ to extract the corresponding anchor hidden feature $z_{s,t} = E_\theta(x_{s,t})$, positive hidden feature $z_{s,t}^{+} = E_\theta(x_{s,t}^{+})$ and negative hidden feature $z_{s,t}^{-} = E_\theta(x_{s,t}^{-})$, respectively. To keep all relation information, we concatenate the feature pairs and input them to an instance relation head $H_\phi(\bullet, \bullet)$ to predict whether they are positive or negative, rather than computing the InfoNCE loss like other contrastive learning methods (e.g., MoCo (He et al. 2020), SimCLR (Chen et al. 2020)). This relation prediction task avoids the demand for massive negative samples in contrastive learning. The pseudo label on whether it is a positive or negative pair can be defined as:

$$\tilde{y}_i = \begin{cases} 1 \text{ if } & p_i^{ins} = H_\phi\left(z_{i,j}, z_{i,j}^{+}\right) \\ 0 \text{ if } & p_i^{ins} = H_\phi\left(z_{i,j}, z_{i,j}^{-}\right) \end{cases} \tag{9.3}$$

where p_i^{ins} is the output of the instance relation head, ϕ denotes its parameter, $z_{i,j}$ denotes an anchor vector, $z_{i,j}^{+}$ and $z_{i,j}^{-}$ are its positive pair and negative pair respectively. Note that

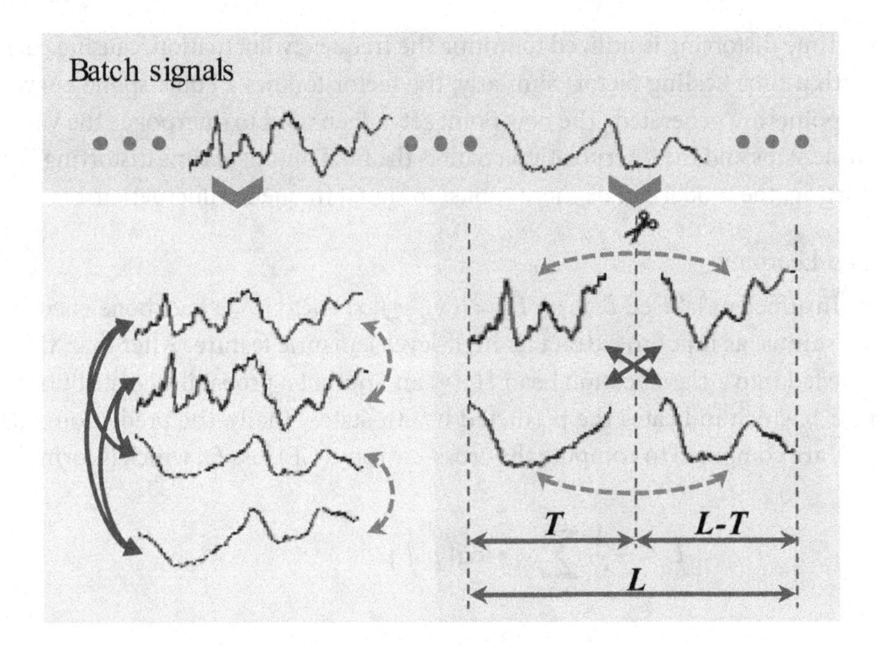

FIGURE 9.3 The relation mining strategy, which selects the positive pairs (dashed double-headed arrows) and negative pairs (solid double-headed arrows) based on (a) the instance relationship of different samples, and (b) the temporal relationship of different segments of signals.

data augmentation is implemented M times, thus for one sample, the total number of its positive pairs is C_M^2. To balance the number of positive and negative pairs, we also select C_M^2 negative pairs randomly. With the auto-generated labels, binary cross-entropy (BCE) is selected as the instance relation loss L_i:

$$L_i = -\frac{2}{BM(M-1)} \sum_{i=1}^{BM(M-1)/2} \tilde{y}_i \bullet \log(p_i^{ins}) + (1-\tilde{y}_i)(1-\log(p_i^{ins})) \tag{9.4}$$

Intra-Temporal Relation Mining

Vibration signals are essentially a type of time series. Their temporal relations can also reveal different fault modes. Therefore, exploring the intra-temporal relation of unlabeled data is effective to extract more discriminative representations. Motivated by next sentence prediction (NSP), which is a pre-trained self-supervised task in BERT (Devlin et al. 2018; Fan et al. 2021), a temporal relation prediction task is adopted to exploit the relationship of different segments in a sample as supervision. Formally, given signal samples $\{x_i \mid x_i \in R^{1\times L}\}_{i=1}^{B}$, we split each sample into two parts $x_i^{past} \in R^{1\times L}$ and $x_i^{future} \in R^{1\times(L-T)}$, where the former T data points refer to its past condition and the latter (L-T) points denote its future condition, as shown in Figure 9.3(b). Based on prior knowledge that the future part of a signal owns similar sequential characteristics to its past part, we randomly select a past segment as the anchor x_i^{anc}. Accordingly, its future counterpart x_i^{pos} is chosen as its positive pair, while another future counterpart x_i^{neg} from a different sample is chosen as its negative pair. Later, the anchor section x_i^{anc}, positive section x_i^{pos} and negative section x_i^{neg} are transmitted to the backbone network, and their hidden representations $z_i^{anc} = E_\theta(x_i^{anc})$, $z_i^{pos} = E_\theta(x_i^{pos})$ and $z_i^{neg} = E_\theta(x_i^{neg})$ are extracted respectively. Like inter-instance relation mining module, a temporal relation head $H_\mu(\bullet, \bullet)$ takes all positive and negative representation pairs as input, and outputs a probability value p_i^{tem} indicating whether the sample pairs are positive or negative. Moreover, pseudo labels $\tilde{y}_i$ depend on the temporal relation of input sample pairs, as illustrated in:

$$\tilde{y}_i = \begin{cases} 1 \ \text{ if } \ p_i^{tem} = H_\mu\left(z_i^{anc}, z_i^{pos}\right) \\ 0 \ \text{ if } \ p_i^{tem} = H_\mu\left(z_i^{anc}, z_i^{neg}\right) \end{cases} \tag{9.5}$$

where μ denotes the parameter of the temporal relation head. Similar to instance relation loss, we choose BCE as the temporal relation loss L_t, which is formulated as:

$$L_t = -\frac{2}{BM(M-1)} \sum_{i=1}^{BM(M-1)/2} \tilde{y}_i \bullet \log\left(p_i^{tem}\right) + (1-\tilde{y}_i)\left(1-\log\left(p_i^{tem}\right)\right) \tag{9.6}$$

where M refers to the applied times of data augmentation, and B is the batch size.

Uncertainty-Based Dynamic Weighting

As mentioned above, two self-supervised losses are designed to explore robust features from unlabeled samples and improve the discriminability of the classifier as auxiliary tasks. It belongs to a challenging multi-task learning problem on how to balance each task and prevent a single task from dominating the optimization process. A naive method is to take the weighted sum of each loss as the total loss. However, manual weight selection is inefficient and sub-optimal. Inappropriate choices of weights may hinder the performance on any individual task. Furthermore, tasks' difficulty varies at different training stages. Constant weighting approach may limit tasks' learning process at a certain stage. To address these problems, we adopted an uncertainty-based dynamic weighting mechanism from (Kendall, Gal, and Cipolla 2018) to avoid model performance degradation.

Generally, the calculation process of the overall network can be expressed as:

$$v = f^W(x) \tag{9.7}$$

$$p(y \mid x) = Softmax(v) \tag{9.8}$$

where W refers to the model parameter, v is the output vector by transmitting the samples x into the model. Later, the vector is normalized through a *Softmax* function to the resulting probability vector $p(y|x)$. Here, $f^W(\bullet)$ refers to the backbone encoder and the projection head. Most classifiers apply *Softmax* function directly, but sometimes scaling the original vector v with a positive random variable σ before *Softmax* function would accelerate the convergence of losses, formulated as:

$$p(y|x,\sigma)=Softmax\left(\frac{1}{\sigma^2}f^W(x)\right) \tag{9.9}$$

where the variance σ^2 is a temperature parameter in Gibbs distribution, which determines the flatness of the final probability distribution. The fluctuation of σ influences the learning difficulty of each loss and reflects the task-specific uncertainty. Following the principled suppose in (Kendall, Gal, and Cipolla 2018), we introduce three learnable standard deviations σ_{sup}, σ_{ins} and σ_{tem} to control the uncertainty of supervised classification task, inter-instance relation prediction task and intra-temporal relation prediction task, respectively. Accordingly, Loss of each task $L(W)$ can be inferred as:

$$\begin{aligned}
L(W)&=-\log p(y=c|x,\sigma)\\
&=-\frac{1}{\sigma^2}f_c^W(x)+\log\sum_{c'}\exp\left(\frac{1}{\sigma^2}f_{c'}^W(x)\right)\\
&=-\frac{1}{\sigma^2}\log p(y|f^W(x))-\log\frac{\left(\sum_{c'}\exp(f_{c'}^W(x))\right)^{\frac{1}{\sigma^2}}}{\sum_{c'}\exp\left(\frac{1}{\sigma^2}f_{c'}^W(x)\right)}\\
&\approx-\frac{1}{\sigma^2}\log p(y|f^W(x))+\log\sigma
\end{aligned} \tag{9.10}$$

where we assume that $\left(\Sigma_{c'}\exp\left(f_c^W(x)\right)\right)^{\frac{1}{\sigma^2}}=\frac{1}{\sigma}\Sigma_{c'}\exp\left(\frac{1}{\sigma^2}f_{c'}^W(x)\right)$ when $\sigma\rightarrow 1$, $f_c^W(x)$ denotes the c'th element of $f^W(x)$. According to mean filed theory, the total conditional probability is equal to the production of the conditional probability of each task:

$$p(y_{sup},y_{ins},y_{tem}|f^W(x),\sigma_{sup},\sigma_{ins},\sigma_{tem})=\prod_{i=sup,ins,tem}p(y_i|f^W(x),\sigma_i) \tag{9.11}$$

Based on (9.10) and (9.11), the dynamic weighted loss is obtained:

$$\begin{aligned}
&L(W,\sigma_{sup},\sigma_{ins},\sigma_{tem})\\
&=\frac{1}{\sigma_{sup}^{\ 2}}L_s(W)+\frac{1}{\sigma_{ins}^{\ 2}}L_i(W)+\frac{1}{\sigma_{tem}^{\ 2}}L_t(W)+\log\sigma_{sup}\sigma_{ins}\sigma_{tem}
\end{aligned} \tag{9.12}$$

From the overall weighted loss, it can be concluded that a large standard deviation σ decreases the weight of the sub-task whereas a small value increases the model's attention to the sub-task. The weight of each task is optimized automatically at each epoch to avoid single task domination, and the last three terms in the equation serve as regularizations to prevent the standard deviation from being too large.

EXPERIMENTAL STUDIES (FOR AEROENGINE BEVEL GEAR FAULT DIAGNOSIS)

In this section, we test three datasets to verify the performance of SSL-based intelligent fault diagnosis framework over other comparative methods.

Datasets

MFPT Bearing Dataset

MFPT bearing datasets were provided by Society for Machinery Failure Prevention Technology (MFPT) (Dataset 2020), which included vibration signals of a health bearing and two fault bearings. The sampling frequency of the health bearing and fault bearing were 97656 Hz and 48828 Hz, respectively. The fault bearings were both operating under 7 different loads. Therefore, there exist 15 health conditions including one health state and 14 fault states. In the following experiment, we select a total of 2560 samples with a length of 1024, including 572 healthy samples and 143 fault samples per fault state.

XJTU-SY Bearing Dataset

XJTU-SY bearing datasets were provided by Xi'an Jiaotong University and Changxing Sumyoung Technology Company (Wang et al. 2018), which included 15 bearing run-to-failure vibration data under 3 working conditions. Every minute, 32768 points were sampled at a frequency of 2.56 kHz. In the following experiment, we collect 1920 samples including 1024 data points, in which each health condition has 128 samples.

Aeroengine Bevel Gear (ABG) Dataset

ABG dataset is a self-designed dataset conducted on an aero-engine lubricating oil accessory test bench to evaluate the performance of bevel gears under different health states. The test bench and the location of sensors are shown in Figure 9.4. Three bevel gear failures

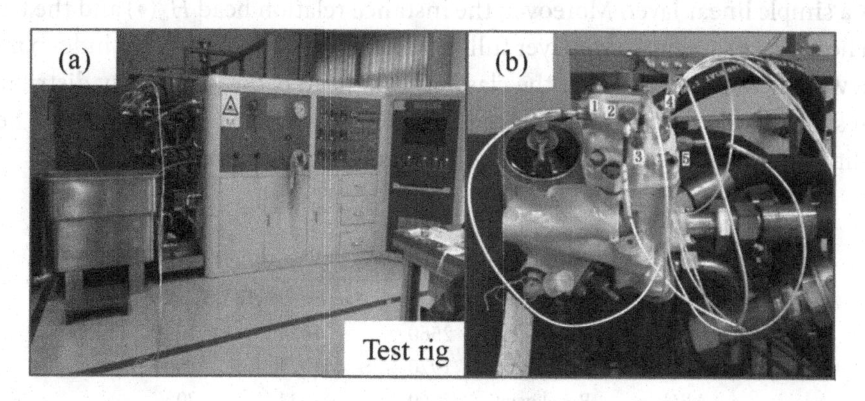

FIGURE 9.4 Aeroengine lubricating oil accessory test bench. (a) Test bench; (b) location of sensors.

FIGURE 9.5 Four states of bevel gears. (a) Health state; (b) tooth surface wear; (c) broken tooth; (d) small end collapse.

are preset on the testbed, that is, tooth surface wear, broken tooth and small end collapse. Together with the normal state, four types of gear health states are shown in Figure 9.5. For each gear health state, vibration signals under five rotating speeds (i.e., 500 rpm, 1000 rpm, 1500 rpm, 2000 rpm, and 3900 rpm) are collected with the sampling frequency of 20 kHz. In the following experiment, a gear state under a certain rotating speed is viewed as a health condition, thus there are 20 health conditions. We generate 1024-length samples without overlapping and select the former 128 samples at each health condition. Details of these three datasets are listed in Table 9.1.

Experimental Description

To avoid test bias, all datasets are firstly split into the training set, validation set and test set with the ratio of 60%:20%:20%. Later, to simulate the label-insufficient setting, we select the label rate α introduced in section II to determine the size of labeled samples. For any health condition with N samples, we only retain labels of random $[\alpha \times N]$ samples and remove labels of other samples. Here, $[\bullet]$ is the floor operator. In later experiments, the labeling rate α is set to 10%, 20%, 40% to evaluate the model performance on different degrees of label scarcity, and 100% to get the upper bound of the performance.

Afterward, all samples are put into the proposed ITSSL, which goes through a backbone encoder and three projection heads. Here, the backbone encoder $E_\theta(\bullet)$ is a four-layer 1D CNN with ReLU activation units and batch normalization (BN). The classification head $H_\varphi(\bullet)$ is a simple linear layer. Moreover, the instance relation head $H_\phi(\bullet)$ and the temporal relation head $H_\mu(\bullet)$ are both two-layer fully connected networks with 256 hidden neurons. It is noteworthy that the output of the classification head is the probability distribution of C health states, while that of the instance relation head and the temporal relation head is a probability value for distinguishing the positive and negative pair.

TABLE 9.1 Details of Three Datasets in Experiment

Dataset	Component	Data Size	Sample Length	Types
MFPT	Bearing	2560	1024	15
XJTU-SY	Bearing	1920	1024	15
ABG	Bevel gear	2560	1024	20

To validate the superiority of the proposed ITSSL, we adopt two supervised learning methods and several SSL methods as comparisons:

1. **Supervised:** A supervised baseline that trains the backbone encoder and the classification head with only labeled data.
2. **Aug_supervised**: An improved supervised method with TAD.
3. **Pseudo-label**: A self-training SSL method that utilizes the pseudo labels of unlabeled data during the training stage.
4. **Π model**: An SSL method using a consensus regularization term to ensure that the predictions of different augmentations are consistent.
5. **MTL**: A co-training SSL method that creates a forecasting task on latent features as regularization.
6. **CR**: A novel SSL method which proposed a contrastive regularization to make features with confident pseudo-labels cluster together.
7. **SLA**: A self-training SSL method that considers the mapping of unlabeled samples to labels as a convex optimization problem and facilitates Sinkhorn algorithm for label assignment.
8. **MoCo**: An advanced contrastive learning method using momentum encoder and memory bank to increase the number of negative pairs. In implementation, it is associated with supervised learning.
9. **SimCLR**: A contrastive learning method which adopts more negative samples and a projection layer. Similarly to MoCo, it is co-trained with supervised learning.

For the fairness of comparison, the model architecture and experiment setups of all methods remain the same. In the training stage, the model is trained with 1000 epochs. An early stopping mechanism with patience of 250 epochs is monitored to prevent overfitting. Besides, Adam optimizer is adopted with a learning rate of 0.01, and the batch size is selected as 128. To alleviate the randomness, we conduct ten random experiments and compute the average test accuracy and standard deviation. The results of all datasets are listed in Table 9.2.

Results and Discussion

Results

As shown in Table 9.2, we can draw the following conclusions: (1) As the label rate decreases, the performance of all methods drops rapidly, which reveals the negative influence of label scarcity problem; (2) the proposed ITSSL can achieve the best accuracy either under label scarcity situations with different label rates or under situations when all data are labeled; (3) Aug_supervised outperforms the supervised baseline across all datasets, which means TAD benefits to improve the diagnostic accuracy; (4) SSL approaches obtain

TABLE 9.2 Experimental Results on Three Datasets

Dataset	MFPT				XJTU-SY				ABG			
Label rate α	10%	20%	40%	100%	10%	20%	40%	100%	10%	20%	40%	100%
Supervised	50.38 ±0.56	56.05 ±0.59	62.91 ±0.62	77.61 ±0.47	68.40 ±0.81	83.15 ±0.73	91.12 ±0.54	98.16 ±0.45	72.72 ±0.79	81.50 ±0.74	88.10 ±0.55	97.03 ±0.37
Aug_ supervised	52.77 ±0.45	60.30 ±0.47	69.90 ±0.51	81.42 ±0.56	72.16 ±0.73	88.45 ±0.66	94.29 ±0.56	98.37 ±0.44	75.69 ±0.68	86.24 ±0.61	91.66 ±0.46	97.16 ±0.31
Pseudo-Label	55.84 ±1.06	61.46 ±0.43	69.64 ±0.79	/	72.05 ±0.69	86.91 ±0.62	95.17 ±0.41	/	75.45 ±0.70	83.81 ±0.52	92.84 ±0.32	/
Π model	60.05 ±0.84	64.76 ±0.73	71.42 ±0.65	/	76.32 ±0.71	88.91 ±0.51	93.09 ±0.28	/	76.96 ±0.64	86.86 ±0.57	91.90 ±0.35	/
MTL	62.17 ±0.65	68.30 ±0.50	75.67 ±0.39	81.98 ±0.38	79.20 ±1.02	91.32 ±0.52	96.99 ±0.30	98.63 ±0.28	76.82 ±0.69	85.54 ±0.62	92.91 ±0.38	97.41 ±0.29
CR	52.37 ±0.85	64.62 ±0.64	70.95 ±0.57	/	69.20 ±0.62	87.47 ±0.58	95.47 ±0.44	/	76.2 ±0.64	90.4 ±0.52	93.2 ±0.43	/
SLA	52.98 ±0.93	63.44 ±0.81	71.34 ±0.66	/	74.53 ±0.75	90.13 ±0.55	95.07 ±0.39	/	80.04 ±0.85	87.60 ±0.66	94.22 ±0.40	/
MoCo	60.28 ±0.73	69.55 ±0.75	75.68 ±0.61	82.99 ±0.58	80.33 ±0.61	**93.47 ±0.63**	95.80 ±0.35	98.43 ±0.26	83.13 ±0.71	90.42 ±0.54	94.18 ±0.44	97.23 ±0.22
SimCLR	61.71 ±0.61	68.58 ±0.54	74.67 ±0.45	81.52 ±0.38	81.43 ±0.70	90.73 ±0.33	95.94 ±0.22	98.59 ±0.19	81.50 ±1.01	89.10 ±0.40	93.80 ±0.33	97.32 ±0.19
ITSSL	**66.17 ±0.78**	**71.72 ±0.74**	**77.31 ±0.43**	**84.47 ±0.45**	**84.73 ±0.91**	92.05 ±0.72	**97.17 ±0.43**	**98.75 ±0.42**	**84.87 ±0.82**	**92.56 ±0.38**	**95.07 ±0.36**	**97.68 ±0.23**

better results than supervised learning methods. Among three self-supervised methods, MoCo and SimCLR achieve excellent accuracy, but there still exists a certain gap when compared with ITSSL. That may attribute to two reasons. First, instance discrimination tasks require large batch size to compare enough negative samples for avoiding model collapse, but the batch size is limited by memory and dataset size. Second, they ignore the temporal relation of vibration signals. (5) From the perspective of datasets, the diagnostic difficulty gradually decreases from MFPT to XJTU-SY to ABG. In the hardest SSL task with only 10% labeled data, the proposed ITSSL improves the diagnostic accuracy over supervised learning baseline by 15.79% on MFPT, 16.33% on XJTU-SY, and 12.15% on ABG, respectively. In the simplest SSL task, i.e., given 40% labeled data on the ABG dataset, the accuracy of the supervised learning method is 88.10%, while our method achieves 95.07%. All results show that ITSSL outperforms supervised methods and other semi-supervised methods, which indicates that the inter-instance and the intra-temporal self-supervised task can effectively capture the underlying features of unlabeled signals and improve the diagnostic performance with few labeled data.

Besides, Table 9.3 presents the training time of ITSSL and the other comparison approaches on three datasets with 20% labeled data. Among all datasets, supervised baseline costs the shortest time since only labeled data participate in calculation. For SSL methods, contrastive learning methods (i.e., MoCo and SimCLR) takes a longer runtime since too many negative pairs are involved. The runtime of ITSSL is comparable to that of other SSL methods, which validates that the proposed method can be applied to other large-scale datasets.

TABLE 9.3 The Training Time (s) of Different Methods on Three Datasets with 20% Labeled Data

Datasets	**MFPT**	**XJTU-SY**	**ABG**
Supervised	11.12	8.21	11.54
Pseudo-label	131.85	99.80	134.64
Π model	343.06	265.89	358.28
MTL	515.21	406.10	535.74
CR	301.49	226.30	308.44
SLA	229.30	175.52	240.68
MoCo	905.94	716.80	917.50
SimCLR	1130.18	856.53	1146.70
ITSSL	827.58	635.73	845.38

Discussions

Ablation Study of Relation Mining Tasks To explore the effect of the inter-instance and intra-temporal relation mining modules, an ablation study is conducted below. Two variants are derived from the proposed ITSSL, i.e., (1) ISSC which only applies inter-instance relation mining module, and (2) TSSC which only utilizes intra-temporal relation mining module. The results of these two variants together with the supervised baseline and the proposed ITSSL on different datasets are shown in Table 9.4.

As can be observed from the results, ISSC, TSSC and ITSSL outperform the supervised baseline by a large extent, which indicates that the combination of self-supervised learning and SSL makes a significant improvement on SSL tasks. Among the three contrastive learning involved methods, TSSC obtains a slightly higher average accuracy than ISSC, which may be attributed to that the temporal features can better reflect the failure modes than inter-instance features. What is more, our method acquires the highest diagnostic accuracy, suggesting that mining both inter-instance relation and intra-temporal relation of unlabeled data can further improve the discriminability of the model.

The Influence of UDW To demonstrate the influence of uncertainty-based dynamic weighting (UDW), we compare it with the traditional fixed weighting approach, namely, the weight of each task is equal and constant during the training process. The comparison

TABLE 9.4 The Results of Ablation Study on Three Datasets

Dataset	**MFPT**				**XJTU-SY**				**ABG**			
Label rate α	10%	20%	40%	100%	10%	20%	40%	100%	10%	20%	40%	100%
Supervised	50.38 ±0.56	56.05 ±0.59	62.91 ±0.62	77.61 ±0.47	68.40 ±0.81	83.15 ±0.73	91.12 ±0.54	98.16 ±0.45	72.72 ±0.79	81.50 ±0.74	88.10 ±0.55	97.03 ±0.37
ISSC	64.47 ±0.84	70.55 ±0.65	76.74 ±0.45	83.10 ±0.43	77.90 ±1.01	90.08 ±0.73	95.97 ±0.39	98.56 ±0.63	78.00 ±0.82	85.90 ±0.53	91.80 ±0.58	97.26 ±0.31
TSSC	65.15 ±0.79	69.76 ±0.84	77.19 ±0.47	83.56 ±0.44	84.16 ±0.93	**94.11 ±0.48**	96.53 ±0.33	98.40 ±0.47	83.90 ±0.76	92.40 ±0.39	94.60 ±0.36	97.32 ±0.42
ITSSL	**66.17 ±0.78**	**71.72 ±0.74**	**77.31 ±0.43**	**84.47 ±0.45**	**84.73 ±0.91**	92.05 ±0.72	**97.17 ±0.43**	**98.75 ±0.42**	**84.87 ±0.82**	**92.56 ±0.38**	**95.07 ±0.36**	**97.68 ±0.33**

TABLE 9.5 The Comparison Results on the MFPT Dataset under Different Label Rates

	Task Weight			Accuracy Under Different Label Rates		
Loss	**Supervised Loss**	**Temporal Relation Loss**	**Instance Relation Loss**	**10%**	**20%**	**40%**
Fixed weighting	0.5	0.5	0	63.46	69.57	73.72
	0.5	0	0.5	60.28	68.77	76.09
	0.33	0.33	0.33	61.66	71.21	75.69
Uncertainty-based dynamic weighting	√	√		65.15	69.76	77.19
	√		√	64.47	70.55	76.74
	√	√	√	**66.17**	**71.72**	**77.31**

results on the MFPT dataset are listed in Table 9.5. It can be observed that when applying the fixed weighting approach, the accuracies of the model that integrates all three losses may be even lower than those using only two losses. It might be owing to a certain loss dominating the training process, thereby preventing other losses from updating. However, when UDW is adopted, the results with both instance relation and temporal relation self-supervised contrasting tasks achieve the best performance. The weight distribution curve of ITSSL during the training process is depicted in Figure 9.6, where the weight of supervised loss drops at the first few epochs and subsequently grows to a stable value. Instead, the weight of instance relation loss increases first and then decreases. And the weight of temporal relation loss gradually drops to a certain value. It indicates that at the initial training stage, the uncertainty of the classification task is larger than relation contrasting tasks and dominates the training process. As the training process goes on, the classification uncertainty decreases, and the model learns a proper weighting scheme.

The Influence of the Split Ratio As mentioned in the intra-temporal relation mining module, the length of the former segment is T and the length of the latter one is L-T. Here we

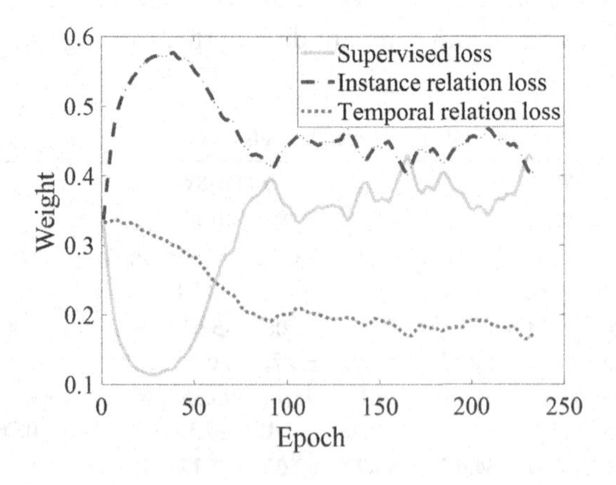

FIGURE 9.6 The weight distribution curve of ITSSL during training process.

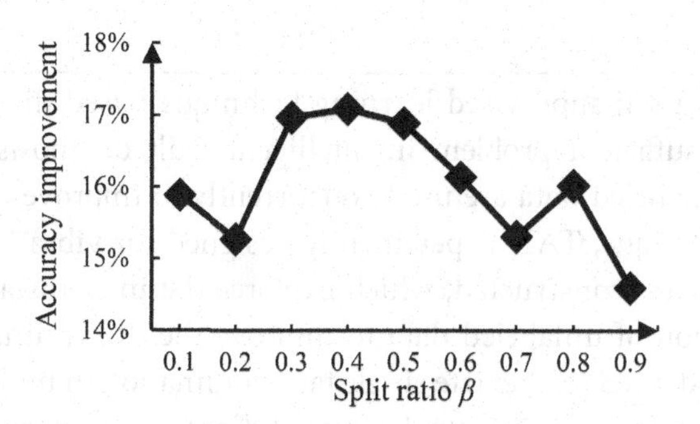

FIGURE 9.7 The accuracy improvement of ITSSL compared with supervised baseline at different split ratios β on MFPT dataset with 40% labeled data.

evaluate the performance of different segment split ratio β, i.e., the ratio of T to L. As shown in Figure 9.7, the black line represents the accuracy improvement of ITSSL compared to the supervised baseline at different split ratios on the MFPT dataset with 40% labeled data. Although the accuracy at each split ratio all outperforms supervised approaches by a large extent, the accuracy improvement is even higher when β is 0.3, 0.4 or 0.5 than the other split ratios. An explanation could be, the contrasting tasks between extremely imbalanced segments are unable to extract the temporal dependency underlying the vibration signals. Generally, the split ratio is selected to 0.5 as a compromise.

Visualization Analysis To illustrate the better feature extracting ability over supervised learning, we visualized the results of the latent features of the backbone encoder on XJTU-SY dataset with only 20% labeled samples, as shown in Figure 9.8, where the left two graphs denote the representations of unlabeled samples and the right two graphs refer to the representations of testing samples. It can be concluded that (1) for supervised learning, the features of various failure modes are mixed, indicating that supervised method owns little discriminability with only few data, (2) as for ITSSL, the self-supervised relation mining tasks can guide unlabeled samples of different failure modes to cluster automatically. Besides, features of testing samples are also pulled away, which makes out-of-distribution fault diagnosis more accurate.

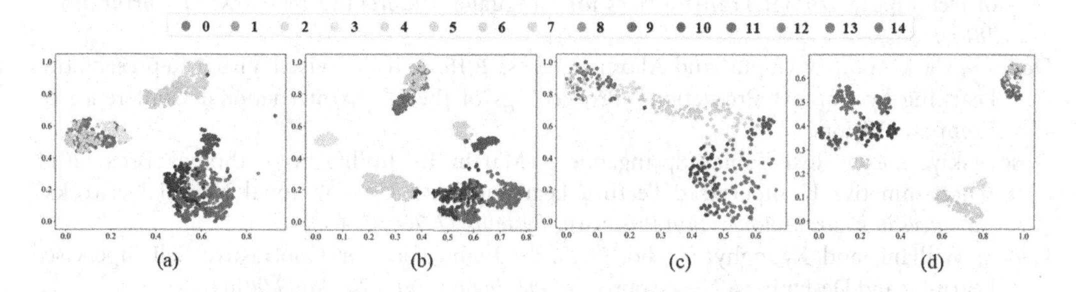

FIGURE 9.8 Visualized features of (a) the unlabeled samples using supervised methods, (b) the unlabeled samples using ITSSL, (c) the testing samples using supervised methods, and (d) the testing samples using ITSSL on XJTU-SY dataset with 20% labeled samples, where 0-14 denote failure modes.

CONCLUSION

This chapter adopts a self-supervised learning technique-based diagnosis framework to address the label-insufficient problem in intelligent fault diagnosis, where few labeled data and massive unlabeled data are used concurrently to improve data efficiency. As an extension of DA technique, TAD is pertinently designed for vibration signals. Then two self-supervised tasks are constructed, which explores the inter-instance relation and the intra-temporal relation of unlabeled data to improve the discriminability of the model. Moreover, UDW is adopted to alleviate single task domination in multi-task learning. Two publicly available datasets and a private dataset validate the superiority of ITSSL to various semi-supervised methods. Ablation studies demonstrate the effectiveness of the temporal relation and instance relation of signals on fault diagnosis. To sum up, self-supervised learning is a useful technique to explore data potentiality, which can be appropriately applied for intelligent tasks in data scarce scenarios.

REFERENCES

Caron, Mathilde, Piotr Bojanowski, Armand Joulin, and Matthijs Douze. 2018. "Deep Clustering for Unsupervised Learning of Visual Features." Proceedings of the European conference on computer vision (ECCV).

Chen, Kaixuan, Lina Yao, Dalin Zhang, Xianzhi Wang, Xiaojun Chang, and Feiping Nie. 2019. "A Semisupervised Recurrent Convolutional Attention Model for Human Activity Recognition." *IEEE Transactions on Neural Networks and Learning Systems* 31 (5): 1747–1756. https://doi.org/10.1109/TNNLS.2019.2927224.

Chen, Ting, Simon Kornblith, Mohammad Norouzi, and Geoffrey Hinton. 2020. "A Simple Framework for Contrastive Learning of Visual Representations." International Conference on Machine Learning.

Choi, Seungdeog, Bilal Akin, Mina M Rahimian, and Hamid A Toliyat. 2010. "Implementation of a Fault-Diagnosis Algorithm for Induction Machines Based on Advanced Digital-Signal-Processing Techniques." *IEEE Transactions on Industrial Electronics* 58 (3): 937–948. https://doi.org/10.1109/TIE.2010.2048837.

Dataset, MFPT. 2020. Society for machinery failure prevention technology.

Deutsch, Jason, and David He. 2017. "Using Deep Learning-Based Approach to Predict Remaining Useful Life of Rotating Components." *IEEE Transactions on Systems, Man, and Cybernetics: Systems* 48 (1): 11–20. https://doi.org/10.1109/TSMC.2017.2697842.

Devlin, Jacob, Ming-Wei Chang, Kenton Lee, and Kristina Toutanova. 2018. "Bert: Pre-training of Deep Bidirectional Transformers for Language Understanding." *arXiv preprint arXiv:1810.04805.*

Doersch, Carl, Abhinav Gupta, and Alexei A Efros. 2015. "Unsupervised Visual Representation Learning by Context Prediction." Proceedings of the IEEE International Conference on Computer Vision.

Dosovitskiy, Alexey, Jost Tobias Springenberg, Martin Riedmiller, and Thomas Brox. 2014. "Discriminative Unsupervised Feature Learning With Convolutional Neural Networks." *Advances in Neural Information Processing Systems* 27: 766–774.

Falcon, William, and Kyunghyun Cho. 2020. "A Framework for Contrastive Self-supervised Learning and Designing a New Approach." *arXiv preprint arXiv:2009.00104.*

Fan, Haoyi, Fengbin Zhang, Ruidong Wang, Xunhua Huang, and Zuoyong Li. 2021. "Semi-supervised Time Series Classification by Temporal Relation Prediction." ICASSP 2021–2021 IEEE International Conference on Acoustics, Speech and Signal Processing (ICASSP).

Gidaris, Spyros, Praveer Singh, and Nikos Komodakis. 2018. "Unsupervised Representation Learning by Predicting Image Rotations." *arXiv preprint arXiv:1803.07728.*

Grill, Jean-Bastien, Florian Strub, Florent Altché, Corentin Tallec, Pierre Richemond, Elena Buchatskaya, Carl Doersch, Bernardo Avila Pires, Zhaohan Guo, and Mohammad Gheshlaghi Azar. 2020. "Bootstrap Your Own Latent-a New Approach to Self-Supervised Learning." *Advances in Neural Information Processing Systems* 33: 21271–21284.

He, Kaiming, Haoqi Fan, Yuxin Wu, Saining Xie, and Ross Girshick. 2020. "Momentum Contrast for Unsupervised Visual Representation Learning." Proceedings of the IEEE/CVF Conference on Computer Vision and Pattern Recognition.

Jawed, Shayan, Josif Grabocka, and Lars Schmidt-Thieme. 2020. "Self-supervised Learning for Semi-supervised Time Series Classification." Advances in Knowledge Discovery and Data Mining: 24th Pacific-Asia Conference, PAKDD 2020, Singapore, May 11–14, 2020, Proceedings, Part I 24.

Jin, Xiaohang, Mingbo Zhao, Tommy WS Chow, and Michael Pecht. 2013. "Motor Bearing Fault Diagnosis Using Trace Ratio Linear Discriminant Analysis." *IEEE Transactions on Industrial Electronics* 61 (5): 2441–2451. https://doi.org/10.1109/TIE.2013.2273471.

Kendall, Alex, Yarin Gal, and Roberto Cipolla. 2018. "Multi-task Learning Using Uncertainty to Weigh Losses for Scene Geometry and Semantics." Proceedings of the IEEE Conference on Computer Vision and Pattern Recognition.

Koch, Gregory, Richard Zemel, and Ruslan Salakhutdinov. 2015. "Siamese Neural Networks for One-Shot Image Recognition." ICML Deep Learning Workshop.

Laine, Samuli, and Timo Aila. 2016. "Temporal Ensembling for Semi-supervised Learning." *arXiv preprint arXiv:1610.02242.*

Lee, Dong-Hyun. 2013. "Pseudo-label: The Simple and Efficient Semi-supervised Learning Method for Deep Neural Networks." Workshop on Challenges in Representation Learning, ICML.

Lee, Doyup, Sungwoong Kim, Ildoo Kim, Yeongjae Cheon, Minsu Cho, and Wook-Shin Han. 2022. "Contrastive Regularization for Semi-supervised Learning." Proceedings of the IEEE/CVF Conference on Computer Vision and Pattern Recognition.

Li, Xiang, Xu Li, and Hui Ma. 2020. "Deep Representation Clustering-Based Fault Diagnosis Method With Unsupervised Data Applied to Rotating Machinery." *Mechanical Systems and Signal Processing* 143: 106825. https://doi.org/10.1016/j.ymssp.2020.106825.

Luo, Minnan, Xiaojun Chang, Liqiang Nie, Yi Yang, Alexander G Hauptmann, and Qinghua Zheng. 2017. "An Adaptive Semisupervised Feature Analysis for Video Semantic Recognition." *IEEE Transactions on Cybernetics* 48 (2): 648–660. https://doi.org/10.1109/TCYB.2017.2647904.

Nie, Xiaoyin, and Gang Xie. 2021. "A Two-Stage Semi-Supervised Learning Framework for Fault Diagnosis of Rotating Machinery." *IEEE Transactions on Instrumentation and Measurement* 70: 1–12. https://doi.org/10.1109/TIM.2021.3091489.

Oord, Aaron van den, Yazhe Li, and Oriol Vinyals. 2018. "Representation learning with contrastive predictive coding." *arXiv preprint arXiv:1807.03748.*

Rasmus, Antti, Mathias Berglund, Mikko Honkala, Harri Valpola, and Tapani Raiko. 2015. "Semi-Supervised Learning With Ladder Networks." *Advances in Neural Information Processing Systems* 28: 3546–3554.

Razavi-Far, Roozbeh, Ehsan Hallaji, Maryam Farajzadeh-Zanjani, Mehrdad Saif, Shahin Hedayati Kia, Humberto Henao, and Gerard-Andre Capolino. 2018. "Information Fusion and Semi-Supervised Deep Learning Scheme for Diagnosing Gear Faults in Induction Machine Systems." *IEEE Transactions on Industrial Electronics* 66 (8): 6331–6342. https://doi.org/10.1109/TIE.2018.2873546.

Sun, Chuang, Meng Ma, Zhibin Zhao, and Xuefeng Chen. 2018. "Sparse Deep Stacking Network for Fault Diagnosis of motor." *IEEE Transactions on Industrial Informatics* 14 (7): 3261–3270. https://doi.org/10.1109/TII.2018.2819674.

Sung, Flood, Yongxin Yang, Li Zhang, Tao Xiang, Philip H. S. Torr, and Timothy M. Hospedales. 2018. "Learning to Compare: Relation Network for Few-Shot Learning." Proceedings of the IEEE Conference on Computer Vision and Pattern Recognition.

Tai, Kai Sheng, Peter D Bailis, and Gregory Valiant. 2021. "Sinkhorn Label Allocation: Semi-supervised Classification Via Annealed Self-Training." International Conference on Machine Learning.

Wang, Biao, Yaguo Lei, Naipeng Li, and Ningbo Li. 2018. "A Hybrid Prognostics Approach for Estimating Remaining Useful Life of Rolling Element Bearings." *IEEE Transactions on Reliability* 69 (1): 401–412. https://doi.org/10.1109/TR.2018.2882682.

Wu, Jingyao, Zhibin Zhao, Chuang Sun, Ruqiang Yan, and Xuefeng Chen. 2021. "Learning from Class-Imbalanced Data With a Model-Agnostic Framework for Machine Intelligent Diagnosis." *Reliability Engineering & System Safety* 216: 107934. https://doi.org/10.1016/j.ress.2021.107934.

Zbontar, Jure, Li Jing, Ishan Misra, Yann LeCun, and Stéphane Deny. 2021. "Barlow Twins: Self-supervised Learning Via Redundancy Reduction." International Conference on Machine Learning.

Zhao, Mingbo, Tommy WS Chow, Peng Tang, Zengfu Wang, Jun Guo, and Moshe Zukerman. 2016. "Route Selection for Cabling Considering Cost Minimization and Earthquake Survivability via a Semi-Supervised Probabilistic Model." *IEEE Transactions on Industrial Informatics* 13 (2): 502–511. https://doi.org/10.1109/TII.2016.2593664.

Zhao, Zhibin, Tianfu Li, Jingyao Wu, Chuang Sun, Shibin Wang, Ruqiang Yan, and Xuefeng Chen. 2020. "Deep Learning Algorithms for Rotating Machinery Intelligent Diagnosis: An Open Source Benchmark Study." *ISA Transactions* 107: 224–255. https://doi.org/10.1016/j.isatra.2020.08.010.

Zhuo, Yue, and Zhiqiang Ge. 2021. "Auxiliary Information-Guided Industrial Data Augmentation for Any-Shot Fault Learning and Diagnosis." *IEEE Transactions on Industrial Informatics* 17 (11): 7535–7545. https://doi.org/10.1109/TII.2021.3053106.

CHAPTER 10

Reinforcement Learning for Intelligent Fault Diagnosis

INTRODUCTION

The planetary gearbox as a critical transmission part of modern mechanical systems has been widely used in the fields of wind turbines, gas turbines, and helicopters because of its advantages of the large transmission ratio, compactness, and strong load-bearing capacity (Chen et al., 2020). However, because planetary gearbox usually operates in harsh environments, its key components, especially the gear and bearing, are frequently subjected to some potential defects such as gear root crack, chipped teeth, broken teeth, pitting, bearing inner race fault, outer race fault, etc. in long-run (Chen et al. 2019). The occurrence of such failures may often cause unexpected downtime of the entire mechanical system, which will result in huge economic losses or even human casualties in accidents (Zhao, Kang, et al. 2017). Thus, it is worth studying about planetary gearbox fault diagnosis methods to ensure the operational safety and reliability of mechanical systems and reduce maintenance costs.

Traditional methods for gearbox fault diagnosis mainly use vibration analysis-based characteristic frequency identification. They utilize signal processing techniques, including short-time Fourier transform (STFT) (Yu, Yu, and Xu 2017), wavelet transform (WT) (Yan, Gao, and Chen 2014), wavelet packet transform (WPT) (Fan and Zuo 2006), empirical mode decomposition (EMD) (Ziani et al. 2019), etc., to extract the failure frequency from vibration signal and realize fault detection. Though this method can detect a fault, it often requires extensive expert knowledge of signal processing and extracts the failure frequency with difficulty under a complex environment (Widodo et al. 2014). With the rise of artificial intelligence (AI), machine learning (ML)-based data-driven approach provides a new way for the development of fault diagnosis. Some ML methods (Ahmed et al. 2019), such as artificial neural networks (Khazaee et al. 2013), support vector machine (SVM) (Peng et al. 2007), extreme learning machine (Yang, Wang, and Wong 2018), etc., have been widely used in fault diagnosis and have shown fruitful results. However, these methods generally need a hand-crafted feature extractor to obtain some time- and frequency-domain features

DOI: 10.1201/9781003474463-12

for fault diagnosis, and the diagnostic accuracy mainly depends on the effectiveness of extracted features (Zhao et al. 2018). Besides, the above classifiers with shallow structures make it difficult to extract abstract features, leading to relatively low accuracy. Especially, with the arrival of the big data era of machinery health monitoring, these methods further show their weakness in dealing with fault diagnosis.

Currently, deep learning (DL), which possesses a strong ability of feature learning, has attracted widespread attention in the field of machine fault diagnosis (Lei et al. 2016). With less manual intervention, these DL-based methods can realize the end-to-end diagnosis and have achieved excellent performance in fault diagnosis.

Besides, because vibration signal often contains strong noise and nonlinear features in practice, some researchers integrated time-frequency analysis (TFA) with DL for fault diagnosis to improve the diagnostic accuracy and robustness (Wang, Yan, and Gao 2017). Through the abovementioned analysis, DL has salient performance on the fault diagnosis of planetary gearbox in contrast to traditional ML methods. However, most methods are standard supervised learning (SL) where the input is directly mapped to a fault type through a classifier in the training process and has strong feedback. Also, their learning ways are static, and unlike the human cognitive mechanism knowledge is learned little by little through interaction with the environment. To a certain extent, these deficiencies reduce the generalization and intelligence level of DL-based fault diagnosis methods. Also, the diagnostic accuracy is influenced if the raw signal is directly used as the model's input because of strong noise and nonlinear characters.

Deep reinforcement learning (DRL), as a breakthrough in AI, fully combines the strong perception of DL with the decision-making advantage of reinforcement learning (RL). However, DRL is seldom used in classification tasks because its solution is aimed at the sequential decision-making problem. For this purpose, Wiering et al. (2011) proposed a classification Markov decision process (CMDP) that defined a standard classification problem as a sequential decision-making problem and an MLP model trained in this framework outperformed a normal MLP trained by backpropagation. Some successful cases (Ding et al. 2019; Ong, Niyato, and Yuen 2020) prompt this chapter to explore a new planetary gearbox's fault diagnosis from the view of DRL to overcome the previously mentioned deficiencies of current fault diagnosis methods.

The rest of the chapter is organized as follows: Section 10.2 gives the problem definition in CMDP. Section 10.3 introduces the theoretical background of synchro-extracting transform (SET) and deep Q network (DQN). Section 10.4 describes the proposed DRL-based fault diagnosis method in detail. Section 10.5 shows the experimental setup and resultant analysis. Finally, Section 10.6 concludes this chapter.

PROBLEM DEFINITION

This section is going to provide background and the problem definition of CMDP.

Problem Description

Fault diagnosis is often deemed as a classification problem. However, DRL generally serves to solve the sequential decision-making problem (e.g., Atari game). Inspired by previous works,

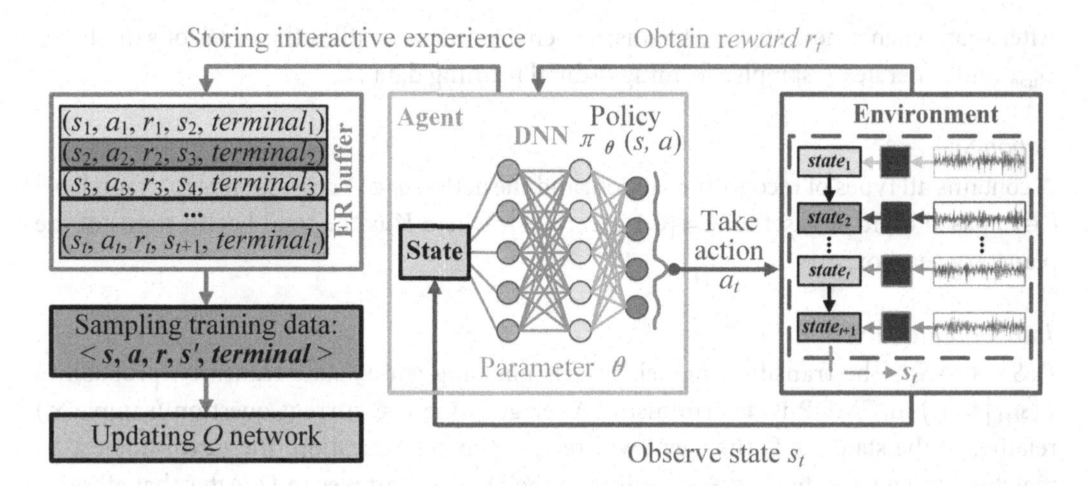

FIGURE 10.1 The overall process of CMDP.

this chapter regards the planetary gearbox's fault diagnosis as a guessing game. Also, a game emulator where fault diagnosis is transformed into a sequential decision-making problem is built, thereby realizing it by using DRL successfully. Suppose that the training dataset that is deemed as a guessing question repertory is $D_{\text{train}} = \{(x_1, l_1), (x_2, l_2), \ldots, (x_n, l_n)\}$ where $x_i \in X$ is the i-th sample and $l_i \in L$ is the i-th label corresponding to the sample x_i. In this game, each round comprises T questions matching training data $D_{\text{train}} = \{(x_1, l_1), (x_2, l_2), \ldots, (x_T, l_T)\}$ that are generated from the training dataset D_{train}, and then an agent guesses these questions sequentially according to the order of samples in D.

In this game, the agent observes a sample at each time and guesses (classifies) its category. Subsequently, the environment returns the agent an immediate reward and the next guessing question (i.e., the next sample), as depicted in Figure 10.1. The game's reward mechanism is defined as that if the agent guesses the category of the sample correctly, a positive reward is awarded, otherwise, a negative reward is given to the agent. The objective of the agent is to maximize accumulated rewards in this game under an optimal behavior policy that is learned from its continual interaction with the environment.

Classification Markov Decision Process

To play the above guessing game using DRL easily, we define this game in a CMDP framework with the tuple $\{S, A, P, R\}$, and form the fault diagnosis problem as a sequential decision-making problem, where S, A, P, R are the state space, action space, transition probability, and reward function, respectively. Considering that the overall process of CMDP is shown in Figure 10.1, each element is described as follows in detail.

State Space S

S consists of all guessing questions in the environment and is determined by training samples. When an episode begins, the agent receives the first sample $x_1 \in X$ as its initial state $s_1 \in S$. Similarly, the state s_t of the environment at each time step matches the sample x_t.

After that, when a new episode begins, the environment shuffles the order of samples in D_{train} and generates T samples forming ordinal training data D.

Action Space A

A contains all types of recognition action, and the action $a_t \in A$ corresponds to a fault label $l_t \in L$. Based on it, A is set to $A = \{0,1,2,\ldots,K-1\}$, where K is the total health status of the planetary gearbox.

Transition Probability P

$P: S \times A \rightarrow S$ is the transition model. In this guessing game, state transition probability $p(s_{t+1} \mid s_t, a_t)$ in CMDP is deterministic. After guessing the current question (sample x_t) relating to the state $s_t \in S$, the agent will receive the next question (the next sample x_{t+1}) matching the next state $s_{t+1} \in S$ according to the order of samples in D. After that all samples in D are guessed sequentially one round game is terminated.

Reward Function R

$R: S \times A \rightarrow R$ denotes a reward r_t obtained by the agent for executing action at the state s_t. The reward function is used to evaluate the success or failure of the agent's actions, thereby guiding the agent to learn the policy π_θ so that it can make relevant actions in given states. In this work, when the agent correctly recognizes a sample x_t a positive reward is given from the environment, otherwise, the agent obtains a negative reward. Thus, the reward function is designed as:

$$R(s_t, a_t, l_t) = \begin{cases} 1, & \text{if } a_t = l_t \\ -1, & \text{otherwise} \end{cases} \tag{10.1}$$

Based on the definitions above, the fault diagnosis problem is formally formulated as a sequential decision-making problem in CMDP, and the objective is to find an optimal recognition policy $\pi_\theta^*: S \rightarrow A$ to maximize accumulated rewards in CMDP. Thus it can be solved by DRL. Also, vibration signal in actual operations often suffers strong noise and nonlinearity because of poor working conditions, which leads to low recognition accuracy if the raw signal is used as the state input directly.

METHODOLOGY

This section is going to introduce the preliminary knowledge of synchro-extracting transform (SET) and DQN, which are utilized in the proposed method based on time-frequency representation (TFR) and DRL.

Synchro-Extracting Transform

SET as a new post-processing step of STFT, only extracts the TF coefficients at the instantaneous frequency (IF) positions to obtain TF maps with more concentrated energy and clear IF curve, thereby enhancing the robustness. Moreover, it outperforms the reassignment method, synchrosqueezing transform (SST), etc., and has low time complexity. Thus, SET is a powerful tool for characterizing inherent features of time-varying nonlinear signal in the time domain and frequency domain simultaneously. This work utilizes SET to transform one-dimensional (1D) fault vibration signals into 2D TF maps suitable for deep convolutional neural network (DCNN) processing and improve diagnostic accuracy, especially for harsh working conditions. SET comprises the following steps:

For a signal $x(u)$, its regular STFT is defined as:

$$G_x(\tau,w)=\int_{-\infty}^{+\infty} x(u)\cdot g(u-\tau)\cdot e^{-iwu}\,du \tag{10.2}$$

where $g(u-\tau)$ indicates a window function (WF) with the center at time τ, w is the frequency, u and τ denote the time parameter.

In considering a phase shift $e^{iw\tau}$, Eq. (10.2) is rewritten as:

$$G_e(\tau,w)=\int_{-\infty}^{+\infty} x(u)\cdot g(u-\tau)\cdot e^{-iw(u-\tau)}\,du \tag{10.3}$$

According to the derivative of $G_e(\tau,w)$, the intrinsic IF $v_G(\tau,w)$ at each TF position can be obtained in:

$$v_G(\tau,w)=\frac{\partial}{\partial\tau}\arg\left(G_e(\tau,w)\right)=i\cdot G_e^{-1}(\tau,w)\cdot G_e^{g'}(\tau,w)+w \tag{10.4}$$

where g' is the derivative of the WF $g(u-\tau)$ regarding time τ.

A new TF plane reconstructed by TF coefficients of $G_e(\tau,w)$ at IF positions, is expressed as:

$$\mathrm{SET}_e(\tau,w)=G_e(\tau,w)\cdot\delta\left(w-v_G(\tau,w)\right) \tag{10.5}$$

where $\mathrm{SET}_e(\tau,w)$ is the TF results of SET, $\delta(\cdot)$ is the Dirac function, and $\delta\left(w-v_G(\tau,w)\right)$ denotes the synchro-extracting operator (SEO). Considering that SEO's real part is utilized in practice, $\mathrm{SEO}(\tau.w)$ is rewritten as:

$$\mathrm{SEO}(\tau,w)=\begin{cases}1, & \left|\mathrm{Re}\left(i\cdot G_e^{-1}(\tau,w)\cdot G_e^{g'}(\tau,w)\right)\right|<\Delta w/2\\ 0, & \text{otherwise}\end{cases} \tag{10.6}$$

where Δw is the frequency interval and it can adjust the SET's frequency resolution, $\mathrm{Re}(\cdot)$ indicates taking the real part.

In this chapter, Δw is set to 1, and the $g(u-\tau)$ is a Gaussian WF with a window length of 128. After the above SET, the 1D time-domain signal is converted to 2D TFR, which will be used as the input of the subsequent model.

Deep Q Network

Deep Q network (DQN), as a representative achievement of DRL, has achieved salient performance in practice when playing Atari games. It mainly integrates deep neural network (DNN) with Q-learning algorithm for autonomously learning the behavior policy π_θ by interaction with the environment, which maps a given state $s \in S$ to an action $a \in A$ such that $a = \pi_\theta(s)$. In the interactive process, the agent is to maximize the cumulated future reward R_t at time t, which is discounted by a factor of γ per time-step and is defined as:

$$R_t = \sum_{t'=t}^{T} \gamma^{t'-t} r_{t'} \tag{10.7}$$

where T indicates the terminated time step, γ is the discount factor. The smaller the value of γ is, the more the agent focuses on the current reward.

For the Q-learning algorithm, its action-value function $Q^\pi(s,a)$ under the policy π is defined as

$$Q^\pi(s,a) = \mathbb{E}_\pi\left[R_t \mid s_t = s, a_t = a; \pi\right] \tag{10.8}$$

where $\mathbb{E}_\pi$ is the expected return under the policy π. The temporal-difference updating of Q-learning is expressed by:

$$Q(s,a) \leftarrow Q(s,a) + \alpha\left(r + \gamma \max_{a'} Q(s',a') - Q(s,a)\right) \tag{10.9}$$

where $Q(s,a)$ indicates the action-value function and α is the learning rate. The optimal action-value function $Q^*(s,a) = \max_\pi Q^\pi(s,a)$ obeys the Bellman equation and is written as:

$$Q^*(s,a) = \mathbb{E}_\pi\left[r + \gamma \max_{a'} Q^*(s',a') \mid s,a\right] \tag{10.10}$$

In DQN, DNN is used to approximate the action-value function $Q^\pi(s,a)$ in the policy π

$$Q(s,a;\theta) \approx Q^\pi(s,a) \tag{10.11}$$

where θ is the parameters of the DNN model.

In the process of the agent's interaction with the environment, ϵ-greedy algorithm, is used to balance "*exploration*" (random action $a \in A$) and "*exploitation*" (action a with $\arg\max_a Q(s,a;\theta)$) and generate a series of experiences defined as $e = (s,a,r,s')$. These experience samples are stored in ER buffer B. Then, ER is utilized to sample experience samples from the ER buffer randomly, thereby updating the DNN model.

Consequently, the mean-square error (MSE) is served as the objective function to update the model's parameters

$$L(\theta)=\mathbb{E}\left[\left(y-Q(s,a;\theta)\right)^2\right] \tag{10.12}$$

where y is the target Q value function and is obtained by:

$$y=r+\gamma\max_{a'}Q(s',a';\theta^-) \tag{10.13}$$

Finally, the gradient descent algorithm is used to update the objective function in (12) to obtain the optimal $Q^*(s,a;\theta)$, thereby inferring the optimal policy $\pi^*=\arg\max_a Q^*(s,a;\theta)$. Accordingly, the gradient updating is calculated by:

$$\nabla_\theta L(\theta)=\mathbb{E}_{e\sim U(B)}\left[\left(y-Q(s,a;\theta)\right)\nabla_\theta Q(s,a;\theta)\right] \tag{10.14}$$

where $e\sim U(B)$ denotes the randomly uniform sampling from ER buffer B. Based on the above DQN, the guessing game defined in Section 10.2 can be solved in the framework of DRL.

DQN-BASED INTELLIGENT FAULT DIAGNOSIS

In this section, a new diagnostic framework integrating SET and DRL is presented in detail, which is mainly composed of data preprocessing, autonomous learning stage, and diagnostic testing. Accordingly, data preprocessing using SET is first introduced. Then the structure of the designed DQN is described and the diagnostic flowchart is shown.

Data Preprocessing

Raw vibration signals from the planetary gearbox are divided into training data and testing data in appropriate proportions. Then, samples are obtained in an overlapping way from the above training or testing data, respectively. The data splitting way is depicted in Figure 10.2.

To utilize time- and frequency-domain information synchronously, this chapter adopts SET to convert 1D vibration sample to 2D TF map, which carries more potential fault information and enhances the robustness than raw signal. According to (10.4) and (10.5), $\text{SET}_e(\tau,w)$ of each vibration sample is obtained and resized to SET_r with a size of $128\times128\times1$, and it is mapped to a TF image (double type) by:

$$P(m,n)=\frac{\text{SET}_r(m,m)-\min(\text{SET}_r)}{\max(\text{SET}_r)-\min(\text{SET}_r)} \tag{10.15}$$

where $P(m,n)\in[0,1]$ is the value at the coordinate point (m,n) of TF image, m and n are all equal to $\{0,1,2,\ldots,127\}$, max and min denote taking maximum and minimum operation,

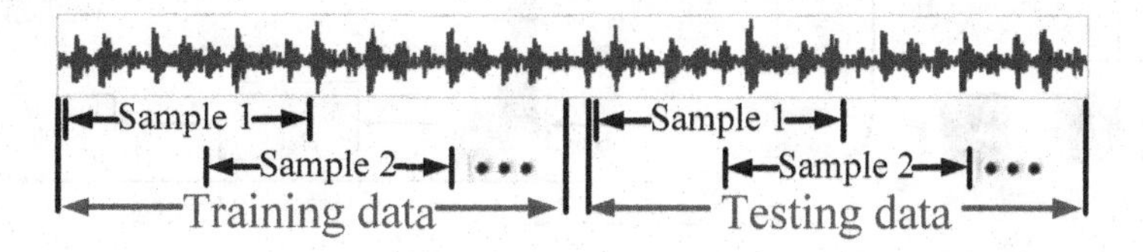

FIGURE 10.2 Data splitting way.

respectively. After the above operations, each image will be regarded as a state observation for the classification agent.

Agent Architecture Design

Considering that the planetary gearbox's fault diagnosis has been converted as a "guessing game" and defined in the CMDP framework as described in Section 10.2, a fault recognition agent based on DQN is presented further in this section. After using SET preprocessing to obtain 2D TF maps, these TF maps will be regarded as the agent's observations for making actions when the agent is playing the guessing game. In this game, the agent will gradually learn the optimal recognition policy by interacting with the environment, thereby realizing fault recognition accurately. The diagnostic framework is depicted in Figure 10.3.

In the method of this chapter, the recognition agent possesses two CNNs with the same structure, where a primary network (Eval-Net) is used to update the current Q value, and another target network (Target-Net) is used to calculate fixed target Q value. The designed CNN structure in Figure 10.3 is described as follows. This CNN has 3 convolutional layers (C-layer), and each C-layer has 8, 12, and 16 filters, respectively. The filter kernel size of the first two C-layer is 3×3, and the final one is set to 5×5. The data padding manner in each C-layer is the "same" mode, and the kernel initializer mode is "*he _normal*". The pool size and stride in Maxpooling2D operation are all set to 2. The activation function of each C-layer and the first fully connected layer (FC1) is a rectified linear unit (ReLU). The second fully connected layer (FC2) corresponds to the output of the model's Q value, and its activation function is linear. The model's state input is the observed TF map with the size of $128 \times 128 \times 1$ obtained by SET. The output of the model is an approximated state-action value $Q(s,a;\theta)$ with K dimensions, and the optimal action corresponds to $\arg\max Q(s,a;\theta)$.

To accelerate the agent's exploration of the environment in the initial learning stage, a dynamic ϵ-greedy algorithm is used to generate the distribution of actions, thereby obtaining

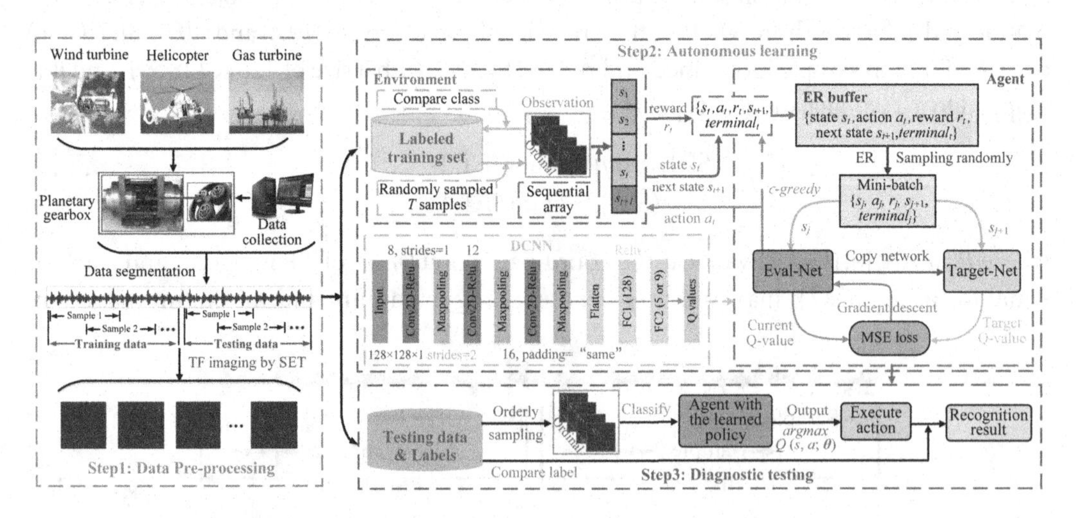

FIGURE 10.3 TFR- and DRL-based intelligent fault diagnosis framework for the planetary gearbox.

various experience samples $e_t = (s_t, a_t, r_t, s_{t+1}, terminal_t)$, which are stored in ER buffer B. The change of parameter ϵ is obtained by:

$$\epsilon = \begin{cases} \epsilon \times \epsilon_{decay}, & \text{if } \epsilon \geq \epsilon_{min} \\ \epsilon_{min}, & \text{otherwise} \end{cases} \tag{10.16}$$

where ϵ_{min} denotes the minimum of ϵ, ϵ_{decay} is the decay factor. In this chapter, the initial ϵ is set to 1, ϵ_{min} is 0.2, ϵ_{decay} is 0.998.

DQN-Based Fault Recognition

Considering the model's training, and CMDP formulated in Section 10.2, a group of T samples is obtained randomly to form a sequential decision-making episode according to the order of these samples when the agent interacts with the environment, as depicted in Figure 10.3. In the process of the agent's autonomous learning, using ER to randomly sample mini-batch experience data $\{e_1, e_2, \ldots, e_{\text{mini-batch}}\}$ from the ER buffer updates the parameters of the agent. Specifically, the updating of Eval-Net is performed through the gradient descent method according to the MSE of the current Q value and target Q value, as illustrated in (11–14), and then Target-Net is updated by copying parameters of Eval-Net after every C step, here C is set to 4. After completing the autonomous learning stage, the agent will make an optimal action $(\arg\max Q(s,a;\theta))$ depending on the learned recognition policy when the testing diagnosis question is given sequentially. Finally, based on Figure 10.3, the flowchart of the proposed method is described especially in Figure 10.4.

In this method, the Adam optimizer with the learning rate of 1e-3 and decay of 1e-4 is used to update the parameters of the agent; the training episode is set to 200; the model's parameter is updated by a minibatch way, and the mini-batch size is set to 32; one round guessing game consists of $T = 128$ questions; the capacity of ER buffer B is 1000. Also, the discount factor γ in (13) is 0.1 because the agent's observations from the guessing game are relatively independent, which means that the agent needs to pay more attention to the current reward in answering a question. It will be also verified in the experiment sections. Also, according to the above settings and the definition of CMDP, we build a simulation environment for training the classification agent.

EXPERIMENTAL STUDIES

This section in turn introduces the Spectra Quest's drivetrain dynamics simulator (DDS) test bench, data description, evaluation indicator, and running system configuration. Finally, results compared with some existing methods are given in detail.

Experimental Description

DDS Testbed and Data Description

Testing Bed The DDS experimental platform is used as the test object to collect various faulty vibration signals under multiple speed-load scenarios, and thereby to verify the

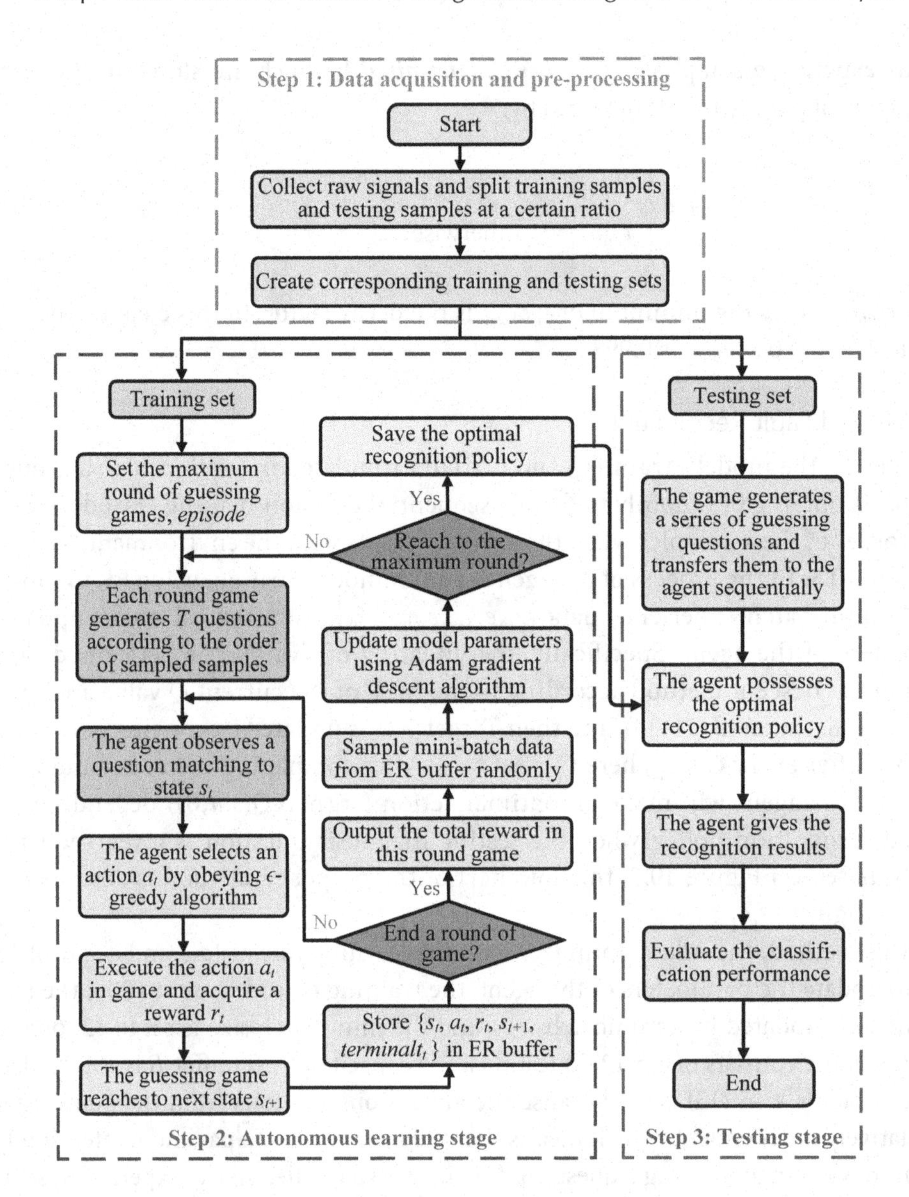

FIGURE 10.4 Flowchart of the methodology in this chapter.

diagnostic performance of the proposed method. This system is mainly composed of a two-stage planetary gearbox, two-stage parallel gearbox, motor, motor controller, brake device, and brake controller, as shown in Figure 10.5. In this chapter, experiments are conducted on the planetary gearbox, whose main parameters are listed in Table 10.1. When the planetary gearbox operates at different conditions, vibration signals are collected by using 608A11 vibrating sensors with a 5120 Hz sampling frequency.

Generally, faults occur in the bearing or gear component. Thus, the experiment simulates nine planetary gearbox's states, including a health status (HEA), four bearing fault types (BWF: ball wear fault; CWF: combo wear fault occurred in an inner race and outer race; IWF: inner race-wear fault; OWF: outer race-wear fault) and four gear fault types

FIGURE 10.5 Planetary gearbox transmission system experimental platform.

(BTF: broken teeth fault; MTF: missing teeth fault; TCF: teeth root crack fault; SWF: surface wear fault), as stated in study in detail, where these fault types and locations are depicted in Figure 10.6. This work mainly analyzes the performance of the proposed method in different speed-load situations. Accordingly, three operating conditions including 20 Hz-0, 30 Hz-2, and 40 Hz-0 are simulated by changing the motor speed and load size. Herein 20 Hz, 30 Hz, and 40 Hz indicate the working speed of a motor, corresponding to 1200 rpm, 1800 rpm, and 2400 rpm, 0 and 2 denote the corresponding load size, corresponding to 0 N·m, 7.2 N·m, respectively.

Data Description In this chapter, three diagnostic subtasks of planetary gearbox, including the fault diagnosis of bearing, gear, and bearing-gear mixture, are carried out. The first two tasks can be regarded as the 5-pattern guessing game, and the last one can be deemed as a 9-pattern guessing game. Each subtask is carried out in three speed-load operating conditions, i.e., 20_0, 30_2, 40_0. Thus, there are nine diagnostic tasks in single-work conditions. In each operating condition, vibration signals of each state are split into 2000 samples in the way depicted in Figure 10.2, where the first 1000 samples are used to train and the last 1000 samples are used to test, and each sample contains 1280 data points. Thus, the first two tasks each have 10,000 samples, and the last task has 18,000 samples.

According to the framework depicted in Figure 10.3, the vibration signal is first converted to the TF image by using SET. As shown in Figure 10.7, a set of TF images of bearing-gear states is obtained by SET under the condition of a 30-2 speed-load situation. From Figure 10.7, these TF maps with more energy concentrated can characterize different states

TABLE 10.1 Main Parameters of Planetary Gearbox

Planetary Gear	Teeth Number Stage 1	Teeth Number Stage 2	Fixed-axle Gear	Teeth Number Stage 1	Teeth Number Stage 2
Sun gear	20	28	Input shaft	29	-
Ring gear	100	100	Jackshaft	100	36
Planet gear	40	36	Output shaft	-	90

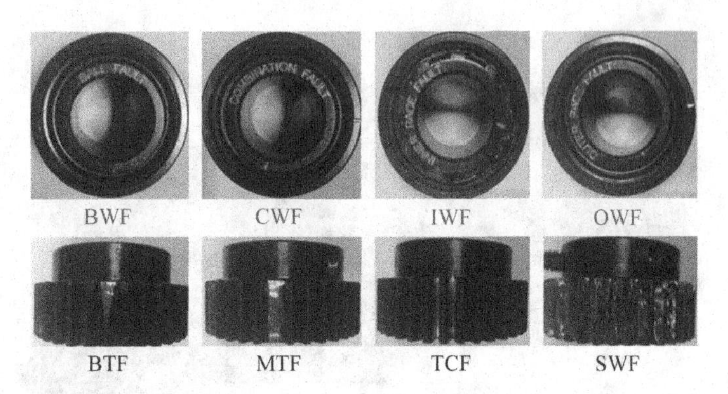

FIGURE 10.6 Fault type descriptions.

of the gearbox to some extent. Then, TF image as the observation in the environment compels the agent to learn a series of recognition policies autonomously, thereby completing the above diagnostic tasks.

Evaluation Indicator and System Configuration

To quantitatively evaluate the diagnostic performance of the proposed approach, we first adopt overall accuracy (OA) to be as an evaluation metric. OA denotes the OA for all classes and is defined as:

$$\mathrm{OA} = \frac{\sum_{i=1}^{M} x_{\mathrm{test,correct}}^{(i)}}{N} \tag{10.17}$$

where $x_{\mathrm{test,correct}}^{(i)}$ is the correctly classified sample size of the i-th category test sample and N is the total number of testing samples.

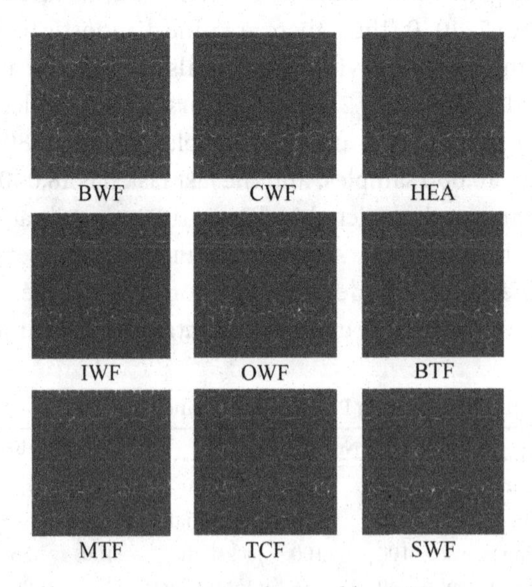

FIGURE 10.7 TF images of bearing-gear states obtained by SET.

Also, the Kappa coefficient is regarded as a statistic of the classification conformance testing (Huang, Nie, and Luo 2019). To further analyze the recognition conformance of the proposed method on each state, the Kappa coefficient (Eugenio and Glass 2004), here named k, is also used as an evaluation metric in this chapter. It is defined as follows:

$$k = \frac{N\sum_{i=1}^{M} x_{\text{test,correct}}^{(i)} - \sum_{i=1}^{M}\left(x_{\text{test,actual}}^{(i)} \times x_{\text{test,predicted}}^{(i)}\right)}{N^2 - \sum_{i=1}^{M}\left(x_{\text{test,actual}}^{(i)} \times x_{\text{test,predicted}}^{(i)}\right)} \tag{10.18}$$

where $x_{\text{test,actual}}^{(i)}$ is the actual total number of the i-th category test sample and $x_{\text{test,predicted}}^{(i)}$ is the predicted total number of the i-th category test sample.

Generally, the parameters of the model are mainly located on the C-layer and FC-layer operations, and the number of parameters (Num) (Chen et al. 2021) is related to the diagnostic time. Thus, the model's trainable Num is used to measure its complexity.

Results and Discussion

Results and Analysis in Single Speed-Load Case

To verify the effectiveness of the algorithm, it is conducted on the above data according to the experimental settings in Section 10.4. When an episode is 200, the change of total reward and loss is statistically analyzed as shown in Figure 10.8. From Figure 10.8, the agent can well learn a series of recognition strategies in training data. Especially in diagnostic tasks of bearing faults, the agent realizes action learning easily. The autonomous learning of the agent is slightly poor in gear or bearing-gear mixture datasets from the change of loss as shown in Figure 10.8 (e) and (f). Additionally, it is easier to learn behavior policy with low speed or no load. When the episode is about 20, the total reward is

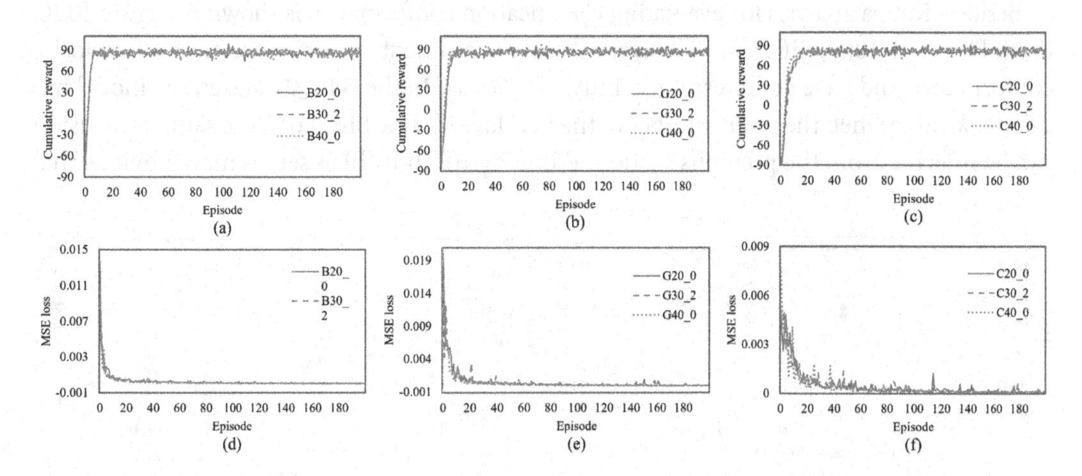

FIGURE 10.8 Change of cumulative reward and loss on bearing, gear, and bearing-gear mixture datasets when respectively executing this algorithm: (a)–(c) Changing curve of reward on the above datasets respectively; (d)–(f) change of training loss on the above datasets, respectively.

relatively constant and revolves around a certain value fluctuation. A reason is that when the training episode exceeds a certain step, the parameter ϵ in the ϵ-greedy algorithm reduces to a constant ϵ_{min} that equals 0.2. It indicates that the agent still makes random "*exploration*" with 20% probability in each time step. Overall, the proposed method has prominent training performance as can be seen from Figure 10.8.

Because the DRL model realizes autonomous learning by only maximizing accumulative expected reward, it may lead to overestimation. Thus three main hyper-parameter effects on results are analyzed in the next section.

Hyper-parameters' Effect on Diagnosis Result

Node Change of FC1 Layer An appropriate model should have as little model complexity as possible while ensuring a certain recognition accuracy. In the designed model as described in Figure 10.3, its trainable Num is mainly located in the FC1 layer. The effect of the node number of the model's FC1 layer on the recognition rate is analyzed. The number of FC1-layer nodes is 32, 64, 128, 256, 512, and 1024, respectively. Besides the above changes, other parameters' setting is the same as described in Section 10.4. To reliably analyze the performance of different FC1 layers, all experiments are independently carried out 5 times in each case. OA, k, and Num are used to evaluate the performance comprehensively. Results of different FC1 layers on the previously mentioned nine tasks are shown in Figures 10.9 and 10.10. Besides, their trainable Num is also compared in Figure 10.11. From Figure 10.9(a), OA on each diagnostic task exceeds 99.5%. It shows that the main body of the designed model is reasonable. When the number of FC1-layer nodes is 128, this model has slightly better overall diagnostic performance and stability, especially in B40_0, G30_2, and C30_2 situations. At this point, the OA of the model in the above situations is 99.94%, 99.92%, and 99.89%, respectively. The average OA of the model on nine datasets reaches 99.9%. From Figure 10.9, it can be also found that the effect of different FC1 layers on B20_0, B30_0, and G40_0 datasets is insignificant.

Besides, Kappa analysis for evaluating classification conformance is shown in Figure 10.10. From Figure 10.10(a), different FC1 layers all have prominent performance on the diagnostic conformance and have good testing stability. Especially in the B40_0 dataset, the model has a better k value when the node number of the FC1 layer is 128. Similarly, the same conclusion can be inferred from the previous section. From Figure 10.10(b), it seems more obvious that

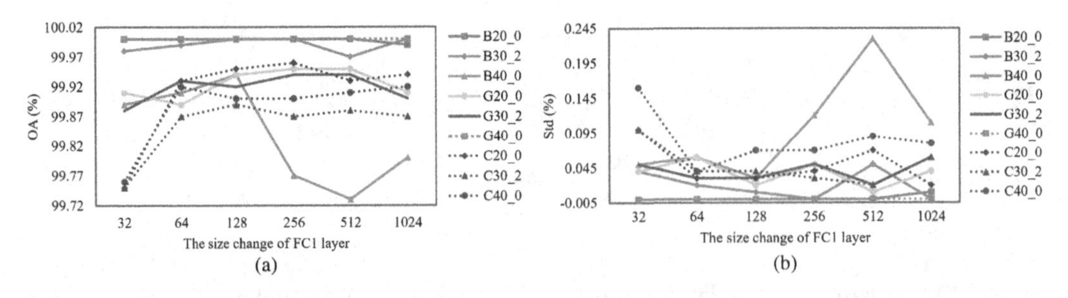

FIGURE 10.9 Results of different FC1 layers. (a) OA analysis of nine working conditions. (b) Std (standard deviation) comparisons.

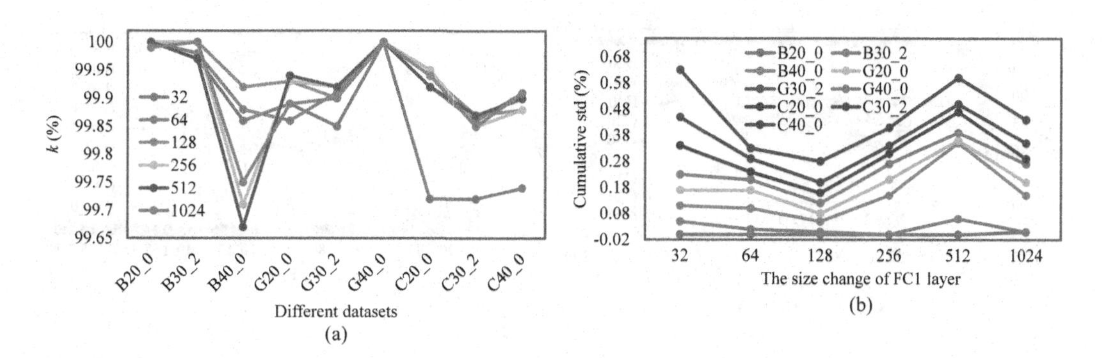

FIGURE 10.10 Kappa analysis of different FC1 layers on multiple datasets. (a) Average k distribution and (b) change of cumulative std on all datasets.

the model has overall better diagnostic stability on nine scenarios when the FC1-layer's node is set to 128. Finally, trainable Num comparisons under different FC1 layers are depicted in Figure 10.11. From Figure 10.11, when the node number of the FC1 layer is 128, the trainable Num of the model is not particularly high by compared with others. Moreover, though the model's learning process is more time-consuming because of incessant self-exploration in the environment for learning the recognition policy, the average diagnostic time of each fault question is about 0.2 ms when the running system is set as described in Section 10.4. Therefore, considering the model size and performance comprehensively, the node number of the model's FC1 layer is set to 128 to realize the fault diagnosis of the planetary gearbox in this chapter.

Change of Training Episode To analyze the effect of parameter episodes on classification accuracy, the proposed method is executed three times on each dataset. For each speed-load case, the change of OA and Std of different training episodes are shown in Figure 10.12. From Figure 10.12, it can be seen that the OA of the proposed method on each sub-dataset is over 99% when the episode arrives at 30. As the value of the episode increases, the overall

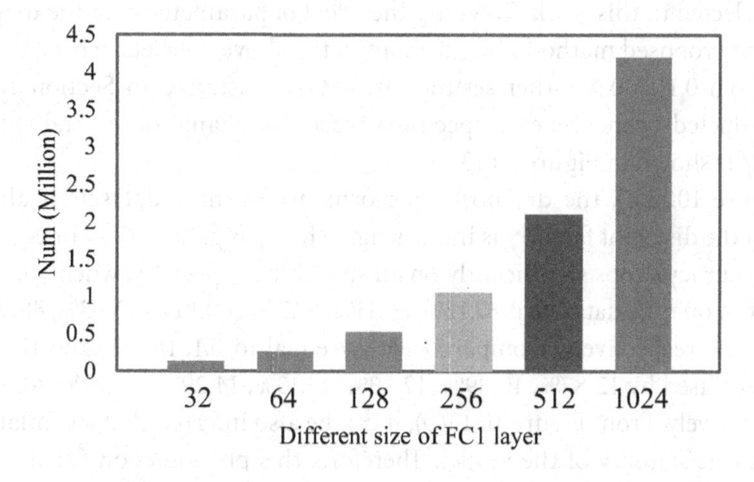

FIGURE 10.11 Trainable num comparisons under different FC1 layers.

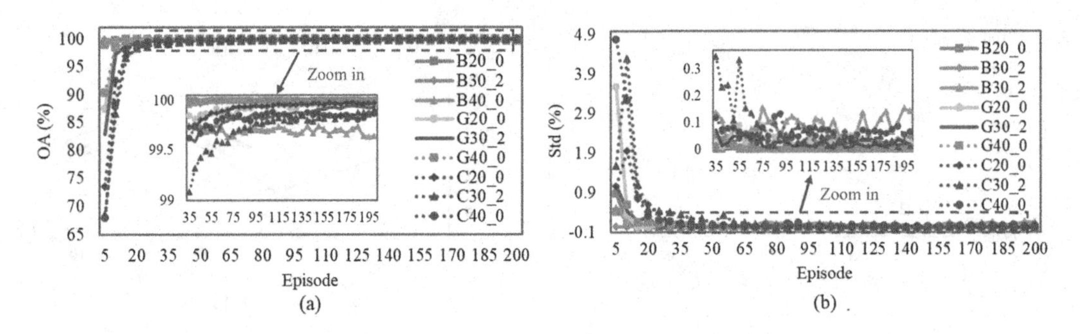

FIGURE 10.12 Classification accuracy change of different training episodes. (a) OA of different training episodes and (b) change of std.

classification accuracy and diagnostic stability are all gradually increasing. When the episode arrives at 100, the OA of each speed-load case is over 99.5%. Compared with other diagnostic tasks, it can be found that fault diagnosis is easier to realize in B20_0, B30_2, and G40_0 conditions. Even in a small episode, this method still has higher accuracy in these diagnostic subtasks. Therefore, the episodes could appropriately be reduced to prevent the agent's overestimation in these subtasks. Since the bearing-gear mixture dataset is a 9-classification task and has more samples, a bigger episode is needed for the agent to explore the experience samples by trial and error. This is the reason why the accuracy is rising slowly in the mixed diagnosis tasks, especially in the C30_2 scenario. Moreover, it can be found that this method performs better in low speed or no load by comparing C20_0 with C30_2 and C40_0 situations. In practical terms, the training episode can be set as a bigger number to increase the agent's interaction with the environment, and thereby improve the accuracy in complex diagnostic tasks.

Change of Discount Factor γ Unlike the actual Atari games, the correlation of adjacent states in the formed "guessing game" is inapparent, thus the classification agent in this "guessing game" cares more about immediate rewards. It is the reason why a small γ value ($\gamma = 0.1$) is selected in this work. To verify the effect of parameter γ on the diagnostic performance, the proposed method is carried out in the above datasets, when the parameters γ changes from 0.1 to 0.9. Other settings are set as illustrated in Section 10.4, and the model is conducted 3 times on each speed-load case. The change of OA and Std in different parameters γ is shown in Figure 10.13.

From Figure 10.13(a), the diagnostic performance on nine datasets is all decreasing overall when the discount factor γ is increasing. When γ is greater than or equal to 0.5, the diagnostic accuracy drops significantly on all sub-tasks. Especially, when γ is equal to 0.9, the model's OA on nine datasets is 87.13%, 89.11%, 87.71%, 84.81%, 85.63%, 88.91%, 79.32%, 84.54%, 84.16%, respectively. Compared with γ equal to 0.1, the diagnostic OA in nine situations decreased by 12.87%, 10.89%, 12.23%, 15.13%, 14.29%, 11.09%, 20.63%, 15.35%, 15.74%, respectively. From Figure 10.13(b), it can be also inferred that a similar conclusion of the diagnostic stability of the model. Therefore, this phenomenon further verified the correctness of the above-said reason why a small γ value ($\gamma = 0.1$) is selected in this work. Importantly, it also gives a reference to how to choose parameters in similar tasks.

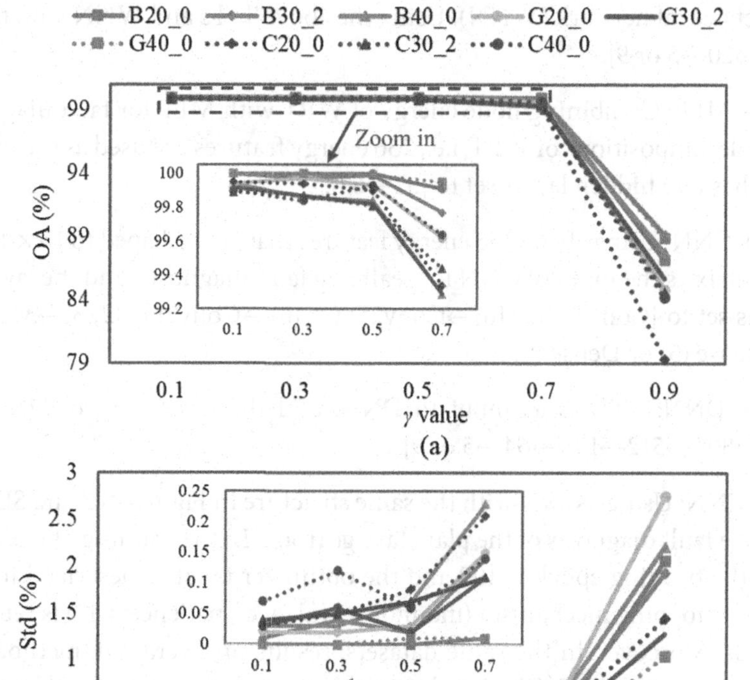

FIGURE 10.13 Results of the proposed method in different parameters γ. (a) Change of OA and (b) stability comparisons of std.

Resultant Comparisons of Different Methods Additionally, several existing methods are also conducted as follows:

a. RTD + MLP: Raw time-domain data (RTD) is as input to MLP, and the layer sizes of MLP are set to [1280→128→5 or 9].

b. Backpropagation (BP): Using 29 time- and frequency-domain features (TFFs) extracted, i.e., mean value, kurtosis, skewness, maximum value, peak-peak value, variance, mean frequency, etc., and BP neural network classifier with the layer size of [29→12→5 or 9] to complete the fault diagnosis.

c. Learning vector quantization (LVQ): Using the LVQ method and the above 29 TFFs to realize fault diagnosis.

d. SVM: Using the above TFFs and SVM classifier with radial basis function (RBF) kernel to complete the fault diagnosis.

e. Probabilistic neural network (PNN): Utilizing the above 29 TFFs and PNN to realize fault diagnosis.

f. Wavelet neural network (WNN): Using the above TFFs and WNN with the structure of [29→20→5 or 9].

g. WPT + MLP: Combining node energy of WPT with MLP for fault diagnosis, where 8-level decompositions of WPT, i.e., 256 energy features are used as the input of MLP, which has two hidden layers set to [32→64].

h. WPT + CNN: Using above 256 energy features that are reshaped to 16 × 16 size of feature matrix as the input of CNN for realizing fault diagnosis, and the layer size of the CNN is set to [Conv2D (3, 16)→Conv2D (3, 16)→Conv2D (32, 3)→Maxpooling2D (2)→Dense (5) or Dense (9)].

i. RTD + DNN: RTD is as input to DNN, and the layer sizes of DNN are set to [1280→800→512→128→64→5 or 9].

j. SET + CNN: Using a CNN with the same structure in Figure 10.3 and SET technique to realize fault diagnosis of the planetary gearbox. In the training of DNN, MLP, and CNN, the training epoch is 100 and the optimizer is set as described in Section IV. The early stopping mechanism (monitor = "val_acc", patience = 10) is set in the above methods. Moreover, in the same datasets, results of several DL methods existed in studies (Shao et al. 2018; Zhao et al. 2017), such as stacked auto-encoder (SAE)-DNN, RNN, LFGRU, method combined CWT with deep transfer learning (DTL) are also listed in this work. These compared results are shown in Table 10.2.

TABLE 10.2 Comparisons of Different Methods (OA %)

	Bearing Dataset			**Gear Dataset**			**Bearing-gear Dataset**			
Method	**20Hz-0**	**30Hz-0**	**40Hz-0**	**20Hz-0**	**30Hz-0**	**40Hz-0**	**20Hz-0**	**30Hz-0**	**40Hz-0**	**AVG**
RTD + MLP	–	88.4	82.64	–	83.4	70.88	–	84.31	72.93	80.42
BP	99.24	99.90	89.66	84.60	83.60	87.16	86.16	81.91	68.88	86.79
LVQ	96.34	98.88	92.12	78.90	79.76	77.54	79.60	78.41	81.50	84.78
SVM	99.76	99.90	87.40	83.14	83.44	82.02	86.89	88.42	81.33	88.03
PNN	97.74	99.68	92.60	85.54	80.30	78.70	82.93	84.92	82.28	87.19
WNN	99.14	99.40	92.580	87.46	87.74	88.12	87.21	87.83	82.58	90.25
WPT + MLP	99.58	99.68	99.38	96.74	95.52	97.80	96.59	96.03	97.73	97.67
RTD + DNN	–	96.89	91.91	–	83.57	92.20	–	92.44	87.13	90.69
SAE + DNN	87.5	92.1	–	92.7	91.9	–	–	–	–	91.05
RNN	92.9	92.0	–	92.3	89.3	–	–	–	–	91.63
GRU	91.2	92.4	–	93.8	90.5	–	–	–	–	91.98
BiGRU	93.0	93.6	–	93.8	90.7	–	–	–	–	92.78
LFGRU	93.2	94.0	–	94.8	95.8	–	–	–	–	94.45
CNN + WPT	99.20	99.22	98.82	94.82	93.80	97.70	96.11	95.74	97.93	97.04
Lfs DCNN	98.90	98.84	–	98.70	94.14	–	98.07	96.40	–	97.51
CWT + DTL	99.94	99.42	–	99.64	99.02	–	99.82	99.31	–	99.53
SET + CNN	99.82	99.80	99.19	99.21	99.13	99.37	99.63	99.54	99.39	99.45
This chapter	100	100	99.94	99.94	99.92	100	99.95	99.89	99.90	99.95

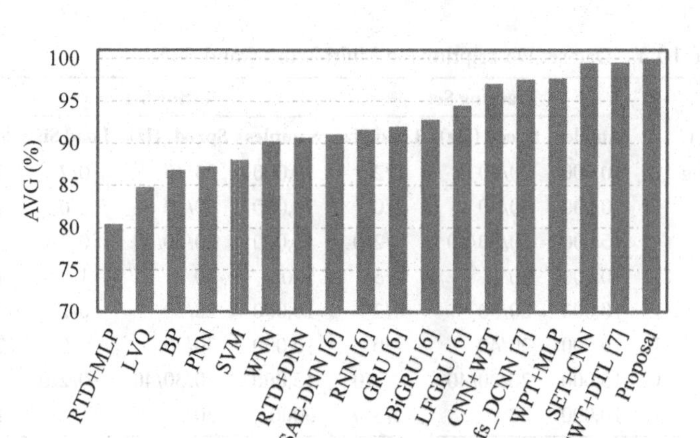

FIGURE 10.14 AVG comparisons of different methods on multiple datasets.

From Table 10.2, it can be found that these methods using feature extraction have better accuracy than the ML-based method that uses raw data as input. In contrast to traditional ML-based fault diagnosis methods, such as BP, SVM, LVQ, PNN, MLP, and WNN, DL-based methods without handcrafted feature extractors all have better fault diagnostic performance. Besides, these methods combining TFA with DL can realize the diagnostic OA of over 90% in multiple datasets. It further proves the advantages of DL in the fault diagnosis of the planetary gearbox. Especially, these three methods, namely CWT + DTL (Shao et al. 2018), SET + CNN, and the proposed method, have more remarkable diagnostic performance.

Compared with others, it can be found that the methods integrating SET and DL or DRL have better diagnostic effectiveness. In the above nine subtasks of the single speed-load case, the OA of the proposed method is 100%, 100%, 99.94%, 99.94%, 99.92%, 100%, 99.95%, 99.89%, and 99.90%, respectively. Especially, in the G30-2 dataset, the overall recognition rate of the proposed method raises 0.9% and 0.79% than CWT + DTL and SET + CNN, respectively. Compared with CNN + SET, the average (AVG) for OA of the proposed method in nine subtasks overall moves up 0.5%. Especially in gear fault diagnosis tasks, the advantage of the proposed method is more obvious. As shown in Figure 10.14, an overall comparison of different methods is obvious from the change of AVG, and it can be further inferred that the proposed method outperforms others.

To sum up, by comparing the results of the above methods in Table 10.2 and Figure 10.14, it can be seen that the proposed method can realize superior diagnostic performance. It also indicates that using SET and DQN can well complete fault diagnosis of the planetary gearbox in a single speed-load case.

Results and Analysis in Multi-Work Conditions

To further evaluate the performance of this method in more complex situations, experiments are carried out in multi-work conditions. The dataset setting and result analysis are as follows.

Dataset Setting Eight tasks in multi-work conditions are carried out. Except for datasets D and H, each subtask contains at least two single speed-load cases, where single speed-load

TABLE 10.3 Dataset Descriptions in Multi-Work Conditions

Dataset		Training Set			Testing Set			Total
		Samples	Speed (Hz)	Load Size	Samples	Speed (Hz)	Load Size	Samples
Bearing	A	10,000	20/30	0/2	10,000	20/30	0/2	20,000
	B	10,000	30/40	2/0	10,000	30/40	2/0	20,000
	C	15,000	20/30/40	0/2/0	15,000	20/30/40	0/2/0	30,000
	D	10,000	30	2/5	5000	30	1	15,000
Gear	E	10,000	20/30	0/2	10,000	20/30	0/2	20,000
	F	10,000	30/40	2/0	10,000	30/40	2/0	20,000
	G	15,000	20/30/40	0/2/0	15,000	20/30/40	0/2/0	30,000
	H	10,000	30	2/5	5000	30	1	15,000

data is mainly from the previously mentioned 20-0, 30-2, and 40-0 data of bearing and gear diagnostic tasks. For datasets D and H, the training data is from the bearing or gear's 30Hz-2 and -5 situations, testing data is from the 30Hz-1 condition. For D and H, the data splitting way is the same as illustrated in Section IV, and each health type has 1000 samples. Detailed settings of the training set and testing set in each situation are introduced in Table 10.3. Each dataset can be deemed as a 5-classification guessing game.

Result and Analysis Results of several methods in multiple working conditions are shown in Figure 10.15. In the method of "RDT + CNN", 1D CNN with the layer size of [Conv1D (32, 7)→maxpooling1D (3)→ Conv1D (64, 3)→maxpooling1D (2)→Conv1D (128, 3)→Maxpooling1D (2)→Dense (5)] is utilized. Other methods' settings are the same as those in the previous section. The model's hyper-parameters are the same as stated in Section 10.4. In eight experiments, OA of the proposed method is 99.99%, 99.78%, 99.94%, 99.99%, 99.82%, 99.84%, 99.77%, and 99.81%, respectively. From Figure 10.15(b), the AVG of these methods is 87.75%, 87.99%, 94.70%, 94.55%, 95.55%, 99.47%, and 99.87%, respectively. Compared with SVM, PNN, MLP, and CNN, the proposed DRL-based method has a better overall diagnosis AVG. In contrast to RTD + CNN, the AVG of the proposed method

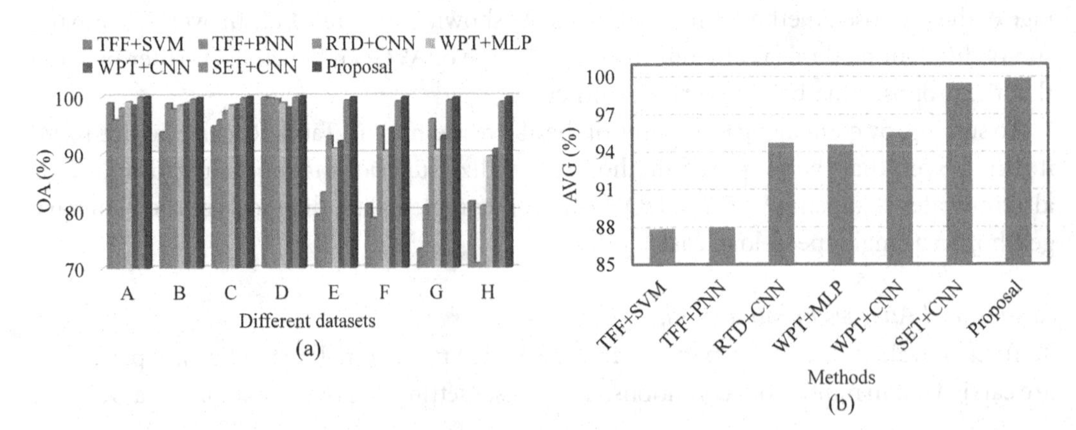

FIGURE 10.15 Results of different methods on multi-work conditions. (a) OA comparisons of different methods and (b) AVG comparisons of eight tasks.

moves up 5.17% in the above eight tasks. Especially in gear datasets E, F, G, and H, this method outperforms others. Also, this method has a better stability. It further proves that this method can still realize planetary gearbox fault diagnosis in multi-work conditions.

From the abovementioned analysis, the proposed method can better identify the planetary gearbox's health states regardless of single or multi-work conditions, especially compared with traditional methods. In addition, DL-based fault diagnosis methods can automatically extract high-quality features that are useful for obtaining high OA of fault diagnosis. Especially for integrating DL and TFR, they further improve the diagnostic accuracy by considering TF information synchronously. It further proves the feasibility and effectiveness of the proposed method. Since the agent needs frequent trial and error to explore the environment when playing the "guessing game", this method requires more time than common methods in the training. But it is not different from SL methods in the testing because of only using the learned policy to make actions without "exploration". Due to the agent's random exploration, it can be regarded as increasing certain noise in the process of model training, thereby improving the model's generalization ability. This may be the reason why the proposed method has a slightly better performance than SET+CNN. Also, it can be found that the agent needs more self-exploration with the increasing of task samples or complexity.

CONCLUSION

To improve the safety and reliability of rotating machinery transmission systems, this chapter transforms planetary gearbox's fault diagnosis as a "guessing game" and presents a new fault recognition method integrating SET and DRL. Moreover, this method is validated under different speed-load conditions on a DDS test platform. Results prove that this approach has better diagnostic accuracy and stability on multiple diagnostic tasks. Even in multi-work conditions, the proposed method still realizes OA over 99.5%. Importantly, it provides a reference for highly accurate and intelligent fault diagnosis for planetary gearbox from a view of DRL.

REFERENCES

Ahmed, Mobyen Uddin, Staffan Brickman, Alexander Dengg, Niklas Fasth, Marko Mihajlovic, and Jacob Norman. 2019. "A Machine Learning Approach to Classify pedestrians' Events Based on IMU and GPS." *International Journal of Artificial Intelligence* 17 (2): 154–167. https://doi.org/10.3390/s17010018.

Chen, Renxiang, Xin Huang, Lixia Yang, Xiangyang Xu, Xia Zhang, and Yong Zhang. 2019. "Intelligent Fault Diagnosis Method of Planetary Gearboxes Based on Convolution Neural Network and Discrete Wavelet Transform." *Computers in Industry* 106: 48–59. https://doi.org/10.1016/j.compind.2018.11.003.

Chen, Shuwen, Hongjuan Ge, Huang Li, Youchao Sun, and Xiaoyan Qian. 2021. "Hierarchical Deep Convolution Neural Networks Based on Transfer Learning for Transformer Rectifier Unit Fault Diagnosis." *Measurement* 167: 108257. https://doi.org/10.1016/j.measurement.2020.108257.

Chen, Siyuan, Yuquan Meng, Haichuan Tang, Yin Tian, Niao He, and Chenhui Shao. 2020. "Robust Deep Learning-Based Diagnosis of Mixed Faults in Rotating Machinery." *IEEE/ASME Transactions on Mechatronics* 25 (5): 2167–2176. https://doi.org/10.1109/TMECH.2020.3007441.

Ding, Yu, Liang Ma, Jian Ma, Mingliang Suo, Laifa Tao, Yujie Cheng, and Chen Lu. 2019. "Intelligent Fault Diagnosis for Rotating Machinery Using Deep Q-Network Based Health State Classification: A Deep Reinforcement Learning Approach." *Advanced Engineering Informatics* 42: 100977. https://doi.org/10.1016/j.aei.2019.100977.

Eugenio, Barbara Di, and Michael Glass. 2004. "The Kappa Statistic: A Second look." *Computational Linguistics* 30 (1): 95–101. https://doi.org/10.1162/089120104773633402.

Fan, Xianfeng, and Ming J Zuo. 2006. "Gearbox Fault Detection Using Hilbert and Wavelet Packet Transform." *Mechanical Systems and Signal Processing* 20 (4): 966–982. https://doi.org/10.1016/j.ymssp.2005.08.032.

Huang, Kui, Wen Nie, and Nianxue Luo. 2019. "Fully Polarized SAR Imagery Classification Based on Deep Reinforcement Learning Method Using Multiple Polarimetric Features." *IEEE Journal of Selected Topics in Applied Earth Observations and Remote Sensing* 12 (10): 3719–3730. https://doi.org/10.1109/JSTARS.2019.2913445.

Khazaee, M, H Ahmadi, M Omid, A Banakar, and A Moosavian. 2013. "Feature-Level Fusion Based on Wavelet Transform and Artificial Neural Network for Fault Diagnosis of Planetary Gearbox Using Acoustic and Vibration Signals." *Insight-Non-Destructive Testing and Condition Monitoring* 55 (6): 323–330. https://doi.org/10.1784/insi.2012.55.6.323.

Lei, Yaguo, Feng Jia, Jing Lin, Saibo Xing, and Steven X Ding. 2016. "An Intelligent Fault Diagnosis Method Using Unsupervised Feature Learning Towards Mechanical Big Data." *IEEE Transactions on Industrial Electronics* 63 (5): 3137–3147. https://doi.org/10.1109/TIE.2016.2519325.

Ong, Kevin Shen Hoong, Dusit Niyato, and Chau Yuen. 2020. "Predictive Maintenance for Edge-Based Sensor Networks: A Deep Reinforcement Learning Approach." 2020 IEEE 6th World Forum on Internet of Things (WF-IoT). https://doi.org/10.1109/WF-IoT48130.2020.9221098

Peng, Lv, Liu Yibing, Ma Qiang, and Wei Yufan. 2007. "Gear Intelligent Fault Diagnosis Based on Support Vector Machines." 2007 Chinese Control Conference. https://doi.org/10.1109/CHICC.2006.4347349.

Shao, Siyu, Stephen McAleer, Ruqiang Yan, and Pierre Baldi. 2018. "Highly Accurate Machine Fault Diagnosis Using Deep Transfer Learning." *IEEE Transactions on Industrial Informatics* 15 (4): 2446–2455. https://doi.org/10.1109/TII.2018.2864759.

Wang, Peng, Ruqiang Yan, and Robert X Gao. 2017. "Virtualization and Deep Recognition for System Fault Classification." *Journal of Manufacturing Systems* 44: 310–316. https://doi.org/10.1016/j.jmsy.2017.04.012.

Wiering, Marco A., Van Hasselt, H., Pietersma, Auke-Dirk, and Schomaker, Lambert. 2011. "Reinforcement Learning Algorithms for Solving Classification Problems." *IEEE Symposium on Adaptive Dynamic Programming and Reinforcement Learning (ADPRL)* 91–96. https://doi.org/10.1109/ADPRL.2011.5967372.

Widodo, Achmad, DP Dewi Widowati, D Satrijo, and Ismoyo Haryanto. 2014. "Vibration Gear Fault Diagnostics Technique Using Wavelet Support Vector Machine." *Applied Mechanics and Materials* 564: 182–188. https://doi.org/10.4028/www.scientific.net/AMM.564.182.

Yan, Ruqiang, Robert X Gao, and Xuefeng Chen. 2014. "Wavelets for Fault Diagnosis of Rotary Machines: A Review With Applications." *Signal Processing* 96: 1–15. https://doi.org/10.1016/j.sigpro.2013.04.015.

Yang, Zhi-Xin, Xian-Bo Wang, and Pak Kin Wong. 2018. "Single and Simultaneous Fault Diagnosis With Application to a Multistage Gearbox: A versatile Dual-ELM Network Approach." *IEEE Transactions on Industrial Informatics* 14 (12): 5245–5255. https://doi.org/10.1109/TII.2018.2817201.

Yu, Gang, Mingjin Yu, and Chuanyan Xu. 2017. "Synchroextracting Transform." *IEEE Transactions on Industrial Electronics* 64 (10): 8042–8054. https://doi.org/10.1109/TIE.2017.2696503.

Zhao, Minghang, Myeongsu Kang, Baoping Tang, and Michael Pecht. 2017. "Deep Residual Networks With Dynamically Weighted Wavelet Coefficients for Fault Diagnosis of Planetary Gearboxes." *IEEE Transactions on Industrial Electronics* 65 (5): 4290–4300. https://doi.org/10.1109/TIE.2017.2762639.

Zhao, Minghang, Myeongsu Kang, Baoping Tang, and Michael Pecht. 2018. "Multiple Wavelet Coefficients Fusion in Deep Residual Networks for Fault Diagnosis." *IEEE Transactions on Industrial Electronics* 66 (6): 4696–4706. https://doi.org/10.1109/TIE.2018.2866050.

Zhao, Rui, Dongzhe Wang, Ruqiang Yan, Kezhi Mao, Fei Shen, and Jinjiang Wang. 2017. "Machine Health Monitoring Using Local Feature-Based Gated Recurrent Unit Networks." *IEEE Transactions on Industrial Electronics* 65 (2): 1539–1548. https://doi.org/10.1109/TIE.2017.2733438.

Ziani, Ridha, Ahmed Hammami, Fakher Chaari, Ahmed Felkaoui, and Mohamed Haddar. 2019. "Gear Fault Diagnosis Under non-Stationary Operating Mode Based on EMD, TKEO, and Shock Detector." *Comptes Rendus Mécanique* 347 (9): 663–675. https://doi.org/10.1016/j.crme.2019.08.003.

Index

Note: Locators in *italics* represent figures and **bold** indicate tables in the text.

E

F

T

U

V

W

X

Printed in the United States
by Baker & Taylor Publisher Services